onservation Trust

Why we are needed:

For thousands of years plants and animals have lived in harmony in our countryside. Increasing pressures on the environment from urban development, transport and extensive agriculture has resulted in the destruction of many important wildlife sites. The Trust believes that areas must be set aside where wildlife can flourish.

Why we need you:

To achieve our aims we need members, money and manpower-and as a charity we need volunteers. By joining the Trust you can gain benefits and enjoyment for yourself as well as helping us achieve our aims.

• reserve

N

ils contact:
e Farm, Hindlip,
K Tel: 0905 754919

The Nature of Worcestershire 1991
has been published in a
Limited Edition
of which this is

Number 347

A list of subscribers
is printed at the
back of this book

THE NATURE OF WORCESTERSHIRE

Marsh warbler. (RMB)

THE NATURE OF WORCESTERSHIRE

THE WILDLIFE AND ECOLOGY
OF THE OLD COUNTY OF WORCESTERSHIRE

WRITTEN AND EDITED BY

G. H. GREEN & BRETT WESTWOOD

with a chapter on the background to the county by

BRIAN ADLAM

Cover illustration by
R. M. BISHOP

Drawings by
R. M. BISHOP, BRETT WESTWOOD
and from the
WORCESTERSHIRE NATURE CONSERVATION TRUST'S FILES

Main photographs by
DAVID M. GREEN & DAVID HOLLIS ARPS

Published with the co-operation and in aid of
THE WORCESTERSHIRE NATURE CONSERVATION TRUST

and with the support of
THE ROYAL SOCIETY FOR NATURE CONSERVATION

DEDICATED TO
CHRISTOPHER CADBURY CBE MA, CONSERVATIONIST EXTRAORDINARY

BARRACUDA BOOKS LIMITED
BUCKINGHAM, ENGLAND
MCMXCI

THE NATURE OF BRITAIN SERIES

PUBLISHED BY BARRACUDA BOOKS LIMITED
BUCKINGHAM, ENGLAND
AND PRINTED BY
DOTESIOS LIMITED
TROWBRIDGE, ENGLAND

COLOUR PLATES PRINTED BY
RAVEN PRINT LIMITED
BRACKLEY, ENGLAND

BOUND BY
WBC BOOKBINDERS LIMITED
MAESTEG, WALES

LITHOGRAPHY BY
FORE COLOUR GRAPHICS LIMITED
HERTFORD, ENGLAND

COLOUR LITHOGRAPHY BY
SOUTH MIDLANDS LITHOPLATES LIMITED
LUTON, ENGLAND

JACKETS PRINTED BY
CHENEY & SONS LIMITED
BANBURY, OXON

TYPE SET IN BASKERVILLE BY
KEY COMPOSITION
NORTHAMPTON, ENGLAND

ISBN 0 86023 487 8

Contents

Colour plates

Acknowledgements

This book is a view from the giant's shoulder of the wildlife and natural history of the old county of Worcestershire. It draws heavily on the observations and records of other people and could not have been written without access to the surveys and records of the Worcestershire Nature Conservation Trust. We are deeply indebted to those who collected and analysed this information, either as part of their working life or as volunteers. We particularly acknowledge the role of the many people who worked for the Trust within the government's various employment schemes, organised mainly through the Manpower Services Commission.

The survey files of the Trust bear many marks made by the hand of John Day. His analyses, made on behalf of the Trust, have been extremely useful to us and his extraordinary knowledge of the wildlife of the county has often provided answers to our questions. Andrew Fraser, John Meiklejohn, Neil Waddingham, Bert Reid, Dave Scott, Roger Maskew and Stuart Brown have answered many queries.

The problem with writing a book like this is that it is selective in what it contains. If we have omitted subjects and failed to mention your favourite places we are sorry to disappoint you. Worcestershire is too rich to be described in a small book.

Abbreviations used in the text

NCC	Nature Conservancy Council
SSSI	Site of Special Scientific Interest
FC	Forestry Commission
WNCT	Worcestershire Nature Conservation Trust

Picture Credits

BA	B. Adlam	JRH	J. R. Hodson
RMB	R. M. Bishop	DH	D. Hollis ARPS
HB	H. Bowkett	WM	Wendy Moule
AWC	A. W. Cundall	TP	T. Parker
DMG	D. M. Green	LS	L. Scott
GHG	G. H. Green	BW	B. Westwood
AH	A. R. Hamblin	WNCT	WNCT files

ENDPAPERS — FRONT: WNCT reserves; BACK: WNCT's new headquarters at Lower Smite Farm. (RMB)

Dedication

This book is dedicated to

CHRISTOPHER CADBURY

in recognition of his immense impact on the conservation of Worcestershire's wildlife.

Christopher Cadbury initiated the founding of the Worcestershire Nature Conservation Trust in 1968. He has remained a powerful influence in the Trust ever since and is President and a Council member.

Through his benefaction he has assisted in the purchase of many nature reserves, including the best in the county, and he has helped in many other ways. Undoubtedly the Trust would not have progressed as well and as far without him. He negotiated space for our first tiny office at Avoncroft College, and persuaded the County Council of the day to fund our first Conservation Officer. Later he gave us office space near his home at Beaconwood. Later still, when the Trust had developed further, he played a leading part in negotiations with the Hereford & Worcester County Council that led the Trust to new larger, headquarters at Lower Smite Farm. Along the way we have received key donations and support from his own Charitable Trust and from other individuals and organisations unable to withstand his persuasive manner. With his support the Trust has established fifty-eight nature reserves, including Upton Warren, a reserve of which he is fond, partly, I think, because he has been able to help create new wetland habitats and to see them visited by many birds new to Worcestershire. He has also been a keen supporter of our dwindling marsh warbler population. When he moved to the Avon Valley a few years ago he quickly planted an osier bed and acquired a wet field in hope of attracting the warblers almost to his back door. I fervently hope they soon return to sing for him!

The thing that pleases me most about Christopher is his passion for land rich in wildlife. Although he fully understands all the ins and outs of conservation politics this seems to be the bedrock on which his conservation achievements lie. He has helped establish reserves for several county Trusts, and overseas, but I hope that he feels that Worcestershire has some of the ones most loved.

This dedication is written in Christopher's eighty-second year. He is twenty-five years my senior and he did not found the Worcester Trust until he was older than I am now! With this example before us we must not (cannot) feel tired or falter in our efforts.

G. H. GREEN
Chairman of Council,
Worcestershire Nature Conservation Trust.

Bibliography and References

This list includes most of the works consulted during preparation of this book. Many of the books and papers contain further references. Books for identifying plants and animals have generally not been included apart from a few important works.

ADLAM, B. H. (ed) (1974) *Worcester and its Region* Worcester Branch of the Geographical Association.
AMPHLETT, M. A. & REA, C. (1909) *The Botany of Worcestershire* Cornish Brothers. Reprinted by EP in 1978.
BARLING, E. (1982) *Birth of a Nature Reserve* Royal Society for Nature Conservation.
BIRKS, J. (1986) *Mink* Anthony Nelson for the Mammal Society.
BROMYARD & DISTRICT LOCAL HISTORY SOCIETY *A Pocketful of Hops* Bromsgrove & District Local History Society.
BROWN, A. (1982) Human impact on the former floodplain woodlands of the Severn. *Brit. Arch. Reps. Int.* Series 146:93-104.
BUCHANAN, K. M. (1944) *The Land of Britain, Report of the Land Utilisation Survey of Britain, Part 63, Worcestershire* Geographical Publications Ltd.
BULLARD, D. W. (1989) *Malvern Hills, A student's guide to the geology of the Malverns* Nature Conservancy Council.
CLAPHAM, A. R., TUTIN, T. G. & MOORE, D. M. (1978) *Flora of the British Isles* Cambridge University Press. Third edition.
CONWAY-JONES, H., (1990) *Working life on Severn & Canal* Alan Sutton.
DARBY, H. C. & TERRETT, I. B. (1971) *The Domesday History of Midland England* Cambridge University Press.
DARLINGTON, A. (1981) *Ecology of Walls* Heinemann Educational Books.
DAVIS, B. N. K. (1983) *Insects on Nettles* Naturalists' Handbooks no. 1. Cambridge University Press.
DEANE, R. J. L. (1989) *Expanded field margins. Their costs to the farmer and benefits to wildlife.* A Report to the Nature Conservancy Council, Kemerton Court, Glos.
FEEK, P. G. (1911) Some notes on the early navigation of the River Avon. In ANDREWS F. B. (ed) *Memorials of Old Worcestershire.* George Allen.
FOWLER, W. WARDE. (1895) *Summer studies of birds and books.* Macmillan.
FRASER, A. J. L. (no date) *Lowland Grasslands of Worcestershire* Information booklet no. 2. Worcestershire Nature Conservation Trust.
FULLER, R. J. & WARREN, M. S. (1990) *Coppiced woodlands: their management for wildlife* Nature Conservancy Council.
GAUT, R. C. (1939) *A History of the Agriculture of Worcestershire and Rural Evolution.* Littlebury.
GODWIN, H. (1975) *History of the British Flora* Cambridge University Press.
GONNER, E. C. K. (1966) *Common Land and Inclosure* Frank Cass.
GREEN, G. H. (1987) Marsh warblers in Worcestershire — personal review. *West Midland Bird Club Annual Report* 54:45-54.
GREEN, J. E. (1982) *A Practical Guide to the Butterflies of Worcestershire.* Worcestershire Nature Conservation Trust.
HADFIELD, M. (1963) Shrawley Wood. *Quarterly Journal of Forestry* 57:35-43.
HAMMOND, C. O. (1977) *The Dragonflies of Great Britain and Ireland* Curwen Books.
HARKER, J. (1989) *Mayflies* Naturalists' Handbooks 13. Richmond.
HARRISON, G. R., DEAN, A. R., RICHARDS, A. J. & SMALLSHIRE, D. (1982) *The Birds of the West Midlands* West Midland Bird Club.
HARRISON, G. & SANKEY, J. (1987) *Where to watch birds in the West Midlands* Christopher Helm.
HARTHAN, J. A. (1938) Some breeding habits of Marsh Warblers in South Worcestershire. *British Birds* 32:230-232.
HARTHAN, J. A. (1947) *The Birds of Worcestershire* Littlebury.
HARLEY, J. B. (1962) *Christopher Greenwood, County Mapmaker* Worcestershire Historical Society.
HASLAM, S. (1978) *River Plants* Cambridge University Press.
HICKIN, N. E. (1981) *Wyre Forest Refreshed* Kenneth Tomkinson.
HICKIN, N. E. (1971) *The Natural History of an English Forest* Hutchinson.
HOOKE, D. (1985) *The Anglo-Saxon Landscape. The Kingdom of the Hwicce* Manchester University Press,
HOWARD, H. ELIOTT (1912) *British Warblers* Porter.
HUMPHREYS, J. (1938) *Studies in Worcestershire History* Cornish.

HURLE, P. (1973) *Beneath the Malvern Hills, A history of the village of Welland.*
HURLE, P. (1978) *Hanley Castle, Heart of Malvern Chase* Phillimore.
HURLE, P. (1984) *The Malvern Hills, A hundred years of conservation* Phillimore.
HURLE, P. (1986) *Malvern Chase* Malvern Museum.
JOHNSON, W. (1981) *Hedge dating and hedgerow flora* Information booklet no. 4. Worcestershire Nature Conservation Trust.
JONES, M. M. (1980) *The Lookers-Out of Worcestershire* Worcestershire Naturalists' Club.
KELSEY, M. G. & GREEN, G. H. (1987) The Decline of Worcestershire's Marsh Warblers. *West Midland Bird Club Annual Report* 54:55-64.
KELSEY, M. G., GREEN, G. H., GARNETT, M. C. & HAYMAN, P. V. (1989) Marsh Warblers in Britain. *British Birds* 82:239-256.
KEMP, R. G. & VICK, G. S. (1983) Notes and observations on *Gomphus vulgatissimus* on the River Severn and River Thames. *J. Brit. Dragonfly Soc.* 1:22-25.
LACK, P. (1986) *The Atlas of wintering birds in Britain and Ireland* British Trust for Ornithology and Irish Wildbird Conservancy, Poyser.
LEES, E. (1867) *The Botany of Worcestershire.* Worcestershire Naturalists' Club.
LLOYD, R. H. (1967) *Bredon Hill and its villages.*
MARCHANT, J. H., HUDSON, R., CARTER S. P. & WHITTINGTON, P. (1990) *Population Trends in British Breeding Birds.* British Trust for Ornithology.
McKNIGHT, H. H. (1975) *The Shell Book of Inland Waterways* David & Charles
MIDDLEMASS, B. — HUNT, J. (1985) *John Corbett Pillar of Salt 1817-1901* Saltway Press, Droitwich.
O'CONNOR, R. J. & SHRUBB, M. (1986) *Farming and Birds* Cambridge University Press.
PACKER, T. H. (1903) *Round Bredon Hill* Crawford.
PAGE, C. N. (1988) *Ferns* Collins.
PERRING, F. H. & WALTERS, S. M. (Eds) (1962) *Atlas of the British Flora* Nelson.
PITT, W. (1813) *General view of the Agriculture of the County of Worcester* David & Charles reprint 1969.
POLLARD, E., HOOPER, M. D. & MOORE, N. W. (1974) *Hedges* Collins.
POTTS, G. R. (1986) *The Partridge* Collins.
PURSEGLOVE, J. (1988) *Taming the Flood* Oxford University Press.
RACKHAM, O. (1980) *Ancient Woodland* Edward Arnold.
RACKHAM, O. (1986) *The History of the Countryside* Dent.
RACKHAM, O. (1990) *Trees and Woodlands in the British Landscape.* Dent. Revised edition.
RATCLIFFE, D. A. (Ed) (1977) *A Nature Conservation Review* Two volumes, Cambridge University Press.
REDFERN, M. (1983) *Insects and Thistles* Natralists' Handbooks no. 4. Cambridge University Press.
RICH, L. (1988) A leader in our time. *Natural World* 23:23-24.
ROLLINS, J. G. (1984) *A History of Redditch* Phillimore.
RUSSELL, R. (1982) *Lost Canals and Waterways in Britain* David & Charles.
SALISBURY, E. (1961) *Weeds and Aliens.* Collins.
SHARROCK, J. T. R. (1976) *The Atlas of Breeding Birds in Britain and Ireland.* British Trust for Ornithology and the Irish Wildbird Conservancy.
SIMMONS, G. E. (1970) The changing face of Shrawley Wood, Worcestershire. *Quarterly Journal of Forestry* 64:60-64.
SNOW, B. & SNOW, D. (1988) *Birds and Berries* Poyser.
THOMAS, J. A. (1986) *Butterflies of the British Isles.* Newnes Country Life Books.
TUCKER, J. J., ZALUCKYI, S. & ALMA, P. J. (1986) *Hartlebury Common. A Social and Natural History* Hereford & Worcester County Council.
TUTIN, T. G. (1980) *Umbellifers of the British Isles.* BSBI Handbook No. 2. Botanical Society of the British Isles.
UNDERDOWN, P. T. (1984) The navigation of the River Severn 1750-1950. *Trans.Worcs. Arch. Soc.* Third series 9:95-100.
VENABLES, A. V. & WILKINS, D. A. (1978) Salt tolerances in pasture grasses., *New Phytol* 80:613-622.
VICTORIA COUNTY HISTORY OF WARWICKSHIRE (1945).
VICTORIA COUNTY HISTORY OF THE COUNTY OF WORCESTER (1901) Originally published by Archibald Constable & Co., Ltd., Reprinted by Dawsons 1971.
WALKER, R. O. (1982) *The Parish of Shrawley.*
WARREN, M. S. & FULLER, R. J. (1990) *Woodland rides and glades: their management for wildlife* Nature Conservancy Council.
WELLS, T., BELLS, S. & FROST, A. (1981) *Creating attractive grasslands using native plant species* Nature Conservancy Council.
YAPP, B. (1962) *Birds and Woods* Oxford University Press.

Introduction

The old county of Worcestershire lies south of the enormous conurbation of the West Midlands. The main body of the county is roughly a rectangle 32 miles north to south, 23 miles east to west, with a twelve-miles long westerly horn extending along the Teme valley to Tenbury Wells. A smaller south-easterly horn at Broadway gathers into Worcestershire a small fragment of the Cotswold escarpment. The total area is approximately 673 square miles.

Historically, ancient Worcestershire was a bigger county, extending further into what is now the southern urban Birmingham, including Northfield, Frankley, Selly Oak, Moseley and Yardley, with Dudley as an outlier surrounded by Staffordshire. Yet another peninsula penetrated Warwickshire to Oldberrow. In the south-east the Broadway prong extended further into the Cotswolds to Blockley, the Fosse Way, and to a large outlier curving north to encompass Shipston-on-Stour and Alderminster, and south to Evenload. In the south the parishes on the south slopes of Bredon Hill were divided between Worcestershire and Gloucestershire with the tower virtually on the county boundary. Tewkesbury and its large enclave to the north, with another west to Forthampton, belonged to Gloucestershire. In the west, Mathon and outlier Edvin Loach were in Worcestershire. The history of the outliers and prongs lies in ancient church ownership, a fascinating subject beyond the scope of this book. The area of the ancient county was 750 square miles. Rationalisation of county boundaries eventually robbed Worcestershire of some fascinating natural history in the depths of Oxfordshire, Warwickshire and the Black Country.

Present-day Worcestershire is mainly bounded by low hills, reaching just over 1,000 feet above sea level at a few points. The fringing hills are all geologically different, forming soils varying from alkaline (basic) on the limestones, to acid on the ancient impervious rocks of the Malverns. Between these extremes much of the central plain of Worcestershire has near neutral, often heavy, clay soils. In the north-west, round Kidderminster, the soils are much more sandy, free-draining, leached and acid. Worcestershire's underlying geology is patchily covered with glacial and water-borne drift deposits with leached acid soils. All this and the complex geology of the West Malvern Hills and the northern Coal Measures leads to a wide variety of soil-types capable of supporting many different plant and animal communities. Additionally the rivers and streams and poor or rapid drainage permit formation of many habitats, ranging from wet marshlands to dry heaths within a few miles of each other. Most of the accepted variety of the English landscape, except for high hills and coast, can be seen within the small county of Worcestershire.

The wonderful natural history of ancient Worcestershire shines through the writings of early naturalists and botanists, especially those of Edwin Lees, who enthusistically leads his readers from one botanical glory to another as he tramps the county during the first part of the nineteenth century: from Cotswold flowers to the old whitty pear in Wyre Forest; from the great marsh at Longdon (influenced by the tides) to the great bog at Moseley (a remnant was recently re-discovered in the heart of Birmingham); from the great limewood at Shrawley to the juniper-clad slopes of the Slads near Craycombe, and from woad on the red marl cliffs near Tewkesbury to deadly nightshade on slopes over the Avon at Cleeve Prior. Even in the 1930s Worcestershire had many remarkable natural attributes and the senior author of this book has hazy remembrances of a pastoral, sleepy, luxuriant county of narrow lands, roadmen cutting verges with sickles, and miles of plum orchards, all doubtless coloured by nostalgia but Lees' quotation from John Clare and applied to Bredon Hill does seem particularly apposite:

'Thou hast fragrant herbs and seed
Which only garden culture need;
Thy Horehound tufts I love them well,
And Ploughman's Spikenard's spicey smell;
Thy thyme, strong scented, 'neath one's feet.
And Marjoram beds so doubly sweet,'

All this has changed. The hand of man has fallen heavily on Worcestershire for the last 150 years and in the last thirty or so it has been heavy indeed. Historically, over 5,000 years man tamed Worcestershire to his own uses and replaced the wildwoods and wetlands with cultivation and habitations. The pattern of hedgerow, field, wood, town and village was finally set by the enclosures, mostly complete by the mid-1800s, though some long before then. That fossillised landscape has been rudely shattered in recent times by intensive agriculture, changed farming needs and practice, vast incrases in population and accompanying urban development, motorways and pollution. We can now chant the nature conservationist's sad litany: 'in the last forty years we have lost 90% of our marshes, 90% of flower-rich meadows, 30% of ancient woodland, over 2,000 miles of hedges, 3,000 ponds, most of our heaths, and so on'. It is all true.

Nevertheless many plants and animals of great interest still lurk in the county, though they are often hard to find. Sometimes a rare plant still grows where Lees found it and sometime one stumbles into a small corner of old England! Marjoram still grows on Bredon Hill. The ten-year New Flora Project, organised by John Day and Roger Maskew, aims to produce a new description of the Flora of Worcestershire and its fieldworkers are finding many unusual plants, including species which have not been seen for decades and several never before recorded in the county. Even on farmland the recent relaxation of pressure to produce even more cereals has led to great surprises as dormant weed seed puts forth flowers and basks in a chemical-free, fire-free, cultivation-free interlude!

Plants are relatively easy to study; much less is known about other groups. Butterflies and dragonflies are fairly easy to identify and there are not many species. But many arthropods, the small insects, spiders and beetles, for example, need 'diligent application' by the observer to identify them, never mind understanding the complexities of their life-cycles and place in the ecology of the county.

In this book we have endeavoured to give an historical perspective to the present natural history of Worcestershire, and to describe the main habitats and main features of the county and some of their interesting plants and animals: rivers, lakes, grasslands, farmland, the fringing hills, woodlands and so on. There is of course, not room to attempt to describe and list everything and we have tried to leaven the list with, we hope, aspects of natural history to catch the reader's attention. We hope to encourage you to go out and enjoy the natural history of Worcestershire, to perhaps become the county's expert on beetles, and to be moved to help and support nature conservation.

Places to Visit

This list contains reserves generally open to everyone, and includes WNCT, Hereford & Worcester county Council, Woodland trust, and a few private properties.

WORCESTERSHIRE NATURE CONSERVATION TRUST RESERVES

The following reserves are generally open to visitors. A full list of reserves is available to members. Further information from WNCT, Lower Smite Farm, Snite Hill, Hindlip, Worcester, WR3 8SZ.

BROADWAY GRAVEL PIT *4 acres, SP 088379. WNCT reserve*
Small, partly flooded, disused gravel pit with scrub and open areas. Small car-park and bird watching hide.
CHANCE WOOD *7 acres, SO 852848. WNCT reserve*
Open broad-leaved woodland.
CLEEVE PRIOR *28 acres, SP 077493 WNCT reserve*
Broad-leaved woodland, scrub, bank of River Avon and limestone grassland.
DEVIL'S SPITTLEFUL & RIFLE RANGE *62 acres and 90 acres, SO 806747 & SO 812749. WNCT reserve*
Contiguous large heathland sites with wooded areas, in Kidderminster.
DUKE OF YORK MEADOW *5 acres, SO 782354. WNCT reserve*
Daffodil meadow near Duke of York pub.
FECKENHAM WYLDE MOOR *28 acres, SP 011603. WNCT reserve*
Wet meadows and marshland. Small pool and marshland. Bird watching hide. Park in Feckenham village: do not drive down track to reserve.
FRED DALE RESERVE *57 acres, SO 775763. WNCT reserve*
Broad-leaved woodland spanning the Dowles Brook in Wyre Forest. Run jointly with West Midland Bird Club.
IPSLEY ALDERS *70 acres, SP 079677. WNCT reserve*
Marsh, pools, wet grasslands and small woods in the Winyates district of Redditch.
KNAPP & PAPERMILL *62 acres, SO 748517. WNCT reserve*
Woods, old orchards and meadows in the Leigh Brook valley, near Alfrick. Small visitors centre.
MONKWOOD *152 acres, SO 803606. WNCT reserve run jointly with British Butterfly Conservation Society*
Broad-leaved woodland with many paths and rides, two miles West of Grimley at Monkwood Green. Small car-park.
NAFFORD ISLAND *SO 941419, WNCT reserve.*
Lock island in the River Avon with scrub and tall herbaceous vegetation. Keep to public footpath.
POOLHAY MEADOWS *7 acres, SO 829308. WNCT reserve*
Lowland flower-rich meadow. Keep to footpath.
RAVENSHILL WOOD *12 acres, SO 739537. Privately owned with some WNCT input*
Mixed woodland with small car-park and Discovery Centre.
TIDDESLEY WOOD *185 acres, SO 929453. WNCT reserve. NE 14 acres private*
Mixed woodland with paths and rides near Pershore. There is a firing range: keep away from zone when red flags are flying.
TRENCH WOOD *106 acres, SO 927586. WNCT reserve with input from British Butterfly Conservation Society*
Broad-leaved woodland. Mature woodland to SW and SE are private.
UPTON WARREN *64 acres. SO 938675. WNCT reserve*
Wetland with lakes, pools and marshes. Inland saltmarsh. Bird watching hides. Visitors should obtain permit from WNCT. Small car-park.
WINDMILL HILL *SO 071474. WNCT reserve*
Limestone grassland and scrub, NE of Evesham.

COUNTRY PARKS ETC.

Further information from Countryside Service, Hereford & Worcester County Council, County Hall, Spetchley Road, Worcestter.
CLENT HILLS COUNTRY PARK
Hills and woodland, Hereford & Worcester County Council, High Harcourt Farm, Walton Hill, Clent, Worcs.

The Pivany Bridge at Knapp & Papermill Reserve. (LS)

FISH HILL PICNIC SITE
Limestone grassland at the top of Broadway Hill.
HARTLEBURY COMMON
Extensive heathland at Stourport-on-Severn. Local Nature Reserve run jointly by Hereford & Worcester County Council, Wyre Forest District Council, with input by Nature Conservancy Council and WNCT.
KINGSFORD COUNTRY PARK
Hereford & Worcester County Council, Blakeshall Lane, Wolverley, Worcs.
WASELEY HILLS COUNTRY PARK
Hereford & Worcester County Council, Gannow Green Lane, Rubery, Worcs.
WORCESTER WOODS COUNTRY PARK
Nunnery Wood and Perry Wood on the outskirts of Worcester, Hereford & Worcester County Council, Countryside Centre, Nunnery Wood, Spetchley Road, Worcester.

MALVERN HILLS CONSERVATORS

MALVERN HILLS
Large tracts of the Malvern Hills and commons are open to visitors.

WOODLAND TRUST
Further information from The Woodland Trust, Autumn Park, Dysart Road, Grantham, NG31 6LL.

PEPPER WOOD *134 acres. SO 940750*
An ancient woodland near Bromsgrove. Once part of the old Forest of Feckenham. Coppice regimes are being re-instated by the WT Community Woodland Project.
WASSELL WOOD *55 acres. SO 794774*
An old oak wood on a hill overlooking the River Severn, near Bewdley, extensively replanted following felling.
UFFMOOR WOOD *212 acres, SO 953811*
Previously owned and managed by Harris Brush Co. Excellent ground flora.

FORESTRY COMMISSION

WYRE FOREST VISITORS CENTRE
Visitor's Centre with many facilities, at the entrance to woodland trails into the Forest. Forestry Commission, Callow Hill, Bewdley, Worcs.

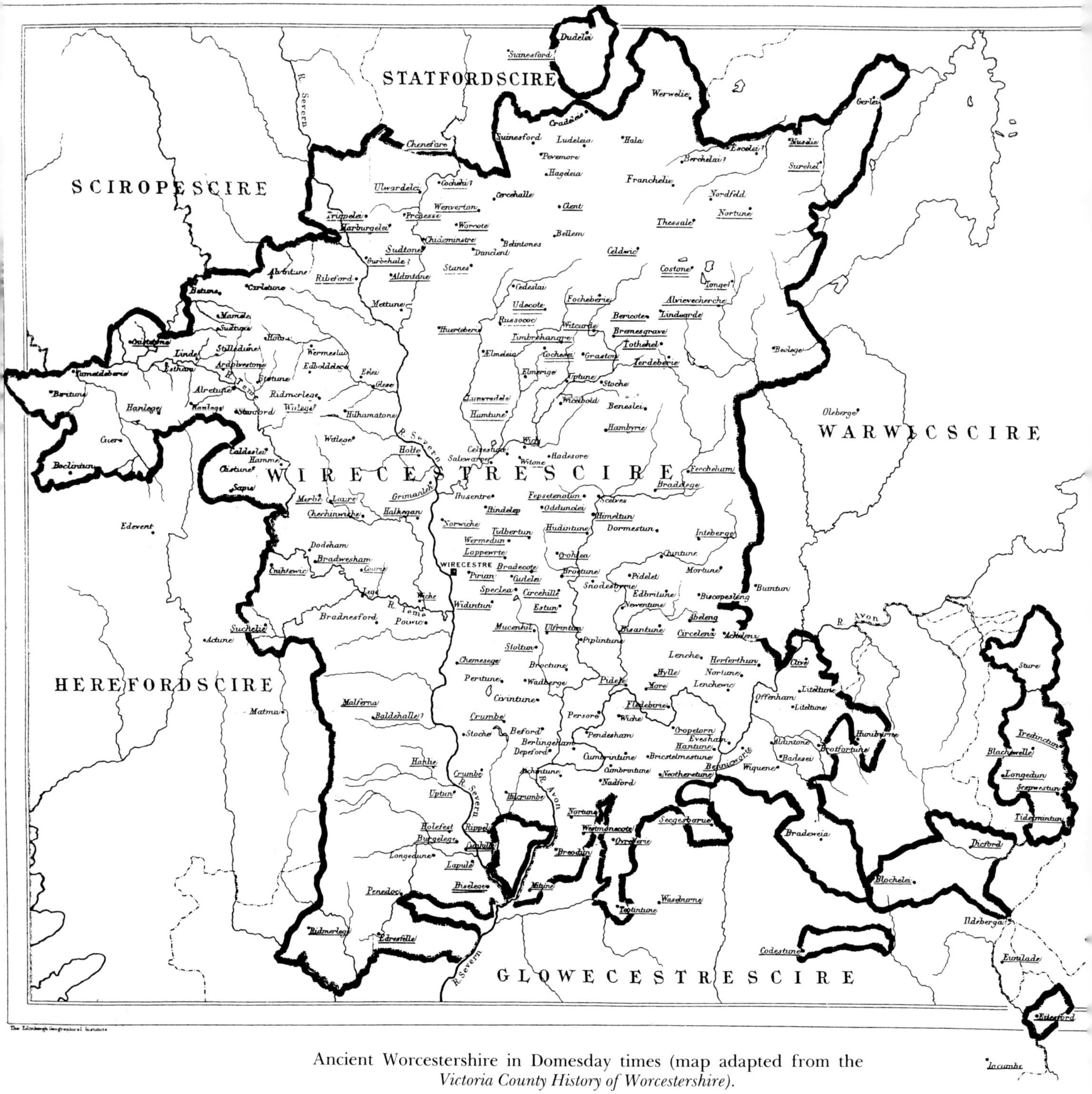

Ancient Worcestershire in Domesday times (map adapted from the *Victoria County History of Worcestershire*).

Background and Origins

by Brian Adlam.

'A marriage, albeit of the shotgun type, has been arranged'. Thus wrote Professor Harry Thorpe of Birmingham University in 1974 when the ancient and historic west midlands counties of Herefordshire and Worcestershire were amalgamated. Yet despite their common Saxon ancestry as Shire counties and the 'desirability' of the merger in the minds of the administrators, the differences between two predominantly rural areas have guaranteed the continuing 'separation'. There is neither physical nor cultural cohesion.

Physiogeographically, Worcestershire is a low-lying county bordered by hills. On the western side is the familiar and stark profile of the Malvern Hills, extending northward along a structural feature known as 'The Malvern Axis', through the Suckley Hills to the Abberley Hills and the Clifton Plateau. On the northern side the Wyre Forest clothes the outermost fringes of the Clee Hills which merge eastwards with the Clent and Lickey Hills, the fringes of the Birmingham Plateau. Less distinctly the eastern margin is marked beyond the shire boundary by the prominent escarpment of the Cotswold Hills, with the noted outliers of Bredon Hill, Dumbleton Hill and Oxenton Hill. Enclosed within this low rim lies the plain of Worcestershire, broken by the three principal river systems of the Severn, Avon and Teme, and opening southwards towards the Vale of Gloucester.

These main regions of Worcestershire correspond broadly to distinct geological eras. The ancient rocks of the Malvern Hills, collectively termed 'Malvernian' and composed of dark crystalline materials, are among the oldest rocks in Britain, with an estimated age of some 620 million years. Against this ancient buttress are compressed the Palaeozoic rocks of the Cambrian, Silurian and Ordovician periods. Dominated by greyish limestones, they occur as a series of irregular and steep-sided hills and valleys, rising to a string of prominent peaks at The Beck, Ankerdine Hill, Berrow Hill, Woodbury Hill and the Abberley Hills.

Beyond and to the north-west, rocks of the Old Red Sandstone and the Coal Measures of the Carboniferous period give way eastwards to Triassic Sandstones, before a brief reappearance of Primary rocks in the Clent and Lickey areas. The eastern side of the county is marked by the southward-trending Ridge Way while, across the River Avon, the Jurassic

Simplified geology of Worcestershire. (BA)

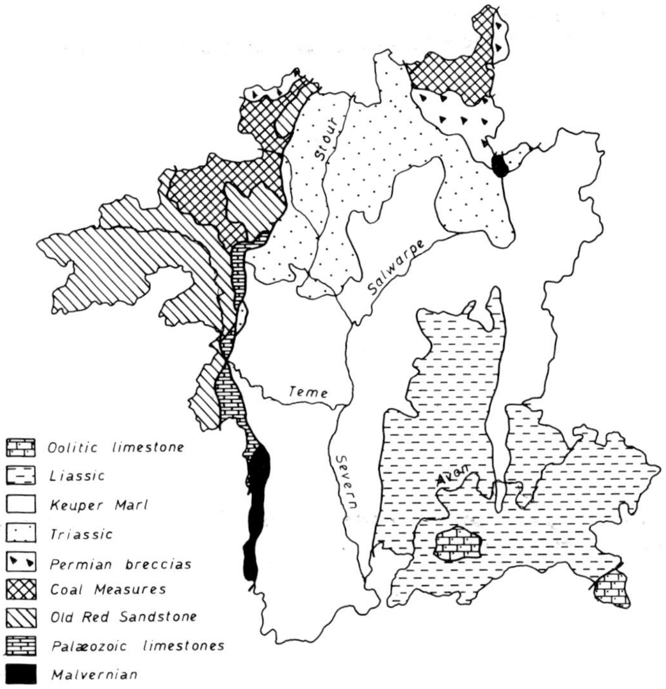

materials of the Cotswold Hills are found. Within this 'inverted horseshoe' of hills, the Plain of Worcestershire is like a great shallow bowl of softer rocks, largely consisting of Keuper Marls and Liassic Clays and with an intermittent cover of 'drift' material derived from the great glaciers and rivers of the past. Great thicknesses of these superficial deposits probably once marked most of the lowland but, since the withdrawal of the icesheets, the area has been washed across by the rivers, swollen by huge volumes of meltwater, so that vestiges only remain, mostly in the form of sand and gravel deposits.

Much of the sand and gravel is of glacial origin, initially deposited on glacial moraines, and later transported, re-shaped and re-deposited along the valley sides and in valley bottoms. For the most part these comprise roughly rounded pebbles of local origin, and with substantial proportions of Triassic materials from the north of the county. However, there are significant quantities of 'foreign' pebbles, derived from distant locations which include the English Lake District, Scotland and Northern Ireland. Such materials could only have been introduced into the West Midlands by glaciers and subsequently re-sorted by fluvio-glacial means.

In economic terms the geology of Worcestershire has not produced great wealth. Coal seams were worked well into the present century in the Halesowen district, and at Clow's Top and Mamble. Perhaps fortunately for us, this coal was used in the Black Country, thus not causing the growth of heavy industrial areas within our shire. Salt is found within the Keuper Marls of the Plain of Worcestershire, and was extensively worked by the Romans. The network of 'salt ways' centring on Droitwich may still be traced, while at Upton Warren the 'Flash Pools', which now constitute one of the county's most important nature reserves, were created by land subsidence following extensive salt (brine) pumping. Elsewhere in the county, the only economic source has been, and still remains, the huge quantities of sand and gravel, particularly the deposits of the Severn Valley, for use in the construction industry.

The rivers of the county are more important than its rocks. Used for centuries as lines of communication, and now as valuable recreational facitities for rowing, sailing and fishing, their present-day role is to supply water, first through the Severn River Board, then the Severn Trent Water Authority (soon to be Severn Trent Water plc).

The drainage pattern of Worcestershire is dominated by the course of the river Severn, the *Sabrina* of Tacitus, which rises among the distant summits of Plynlimon in central Wales. Flowing through parts of Denbighshire, Montgomeryshire (Dyfed and Powys) and Shropshire, it crosses the boundary of Worcestershire near Bewdley, traverses the county north to south and continues through Gloucestershire to the sea. Within Worcestershire it is joined by its two principal tributaries: the Teme, which also has its source in mid-Wales, at Worcester, and the Warwickshire Avon just north of Tewkesbury. With the exception of the Avon, one of the most striking components of the drainage of Worcestershire is the number of watercourses which have a strong north-to-south orientation; for example the Severn itself, parts of the Teme, the brooks along the Malvern axis and such streams as the Piddle and Bow Brook on the west side of the Vale of Evesham. With the exception of the streams along the Malvern Axis, there is little geological variation to explain this 'parallelism' of streams so, in the absence of structural influences, an alternative explanation must be sought. It seems probable that the pattern may have been created at a time when huge quantities of meltwater were flowing away from the edges of the great ice sheets which lay to the north in the Pleistocene period. The gradual diminution of these glacial torrents seems to have resulted in the emergent streams adopting to a large degree those meltwater channels.

There are two curious 'discordances' in the drainage pattern, one the course of the Teme at Knightwick Bridge, the other on the Leigh Brook in the vicinity of Longley Green. In both cases the rivers flow directly across the north-to-south line of the Malvern axis, a massive structural feature which might be expected to dictate a similar orientation for its rivers. Again, the glacial history of the county provides the explanation.

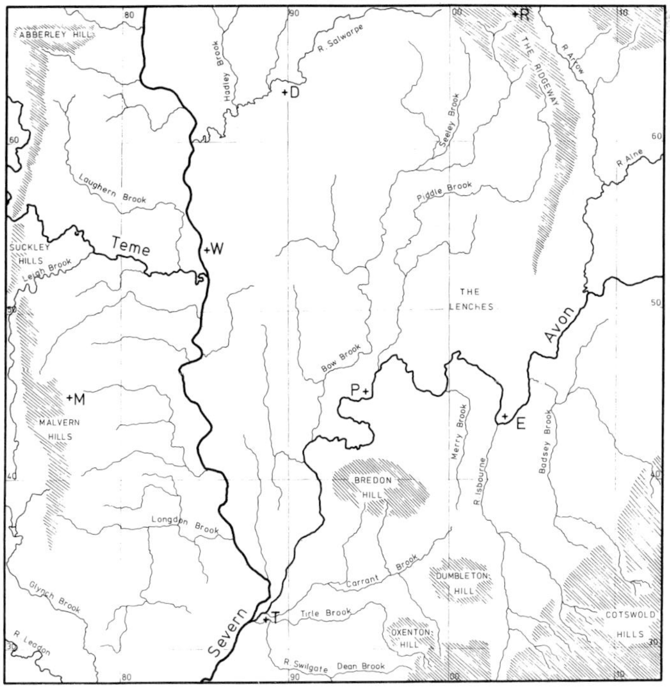

Principal drainage lines in Worcestershire. (BA)

Although Worcestershire lies well within the limits of the maximum extent of the glacial ice in Britain during the Wolstonian period some 120,000-70,000 BP (Before Present), there is a marked paucity of glacial deposits. Landscape evolution has therefore to be deduced from the form and shape of the surface, and from the random and scattered deposits which still exist. Certainly during the last glacial phase of the Quaternary, the Devensian period of some 35,000-12,000 BP, there was little ice here, but plenty of water. Besides apparently controlling the course of present-day rivers and removing huge quantities of glacial morainic debris, the meltwater torrents appear to have had profound effects upon the principal rivers.

Prior to the glacial advances it is likely that the Avon flowed north-eastwards, opposite to the present direction, as a tributary to the Trent river system, while the upper or Welsh Severn seems likely to have discharged into the Irish Sea, probably as a tributary of the river Dee. Worcestershire would then have been drained only by a small stream which rose somewhere near Wolverhampton, picking up tributaries from east and west during its course south through Worcestershire. The Teme at that time was almost certainly part of the river Wye system. Professor F. W. Shotton, in 1953, described how the Avon was completely reversed in its direction of flow by ice which advanced up the Trent valley system from the north-east. That ice formed an 'ice dam' across which the river could not escape, causing the formation of Lake Harrison, which eventually over-spilled across its own headwaters area into the little Severn valley of the time. Professor L. J. Wills, in 1924, showed how ice in the plain of Shropshire had a similar effect on the Welsh Severn, pounding up a series of lakes behind glacial dams until they spilled out over their enclosing rims, in this case creating the famous Ironbridge Gorge. P. Cross and B. H. Adlam, in 1971, finally showed how yet another body of ice pounded up the Welsh Teme within the Wye system, creating yet another lake in the vicinity of Tenbury Wells (Lake Woofferton) which eventually over-spilled at Stockton-on-Teme. According to the various dating techniques applied, the Avon

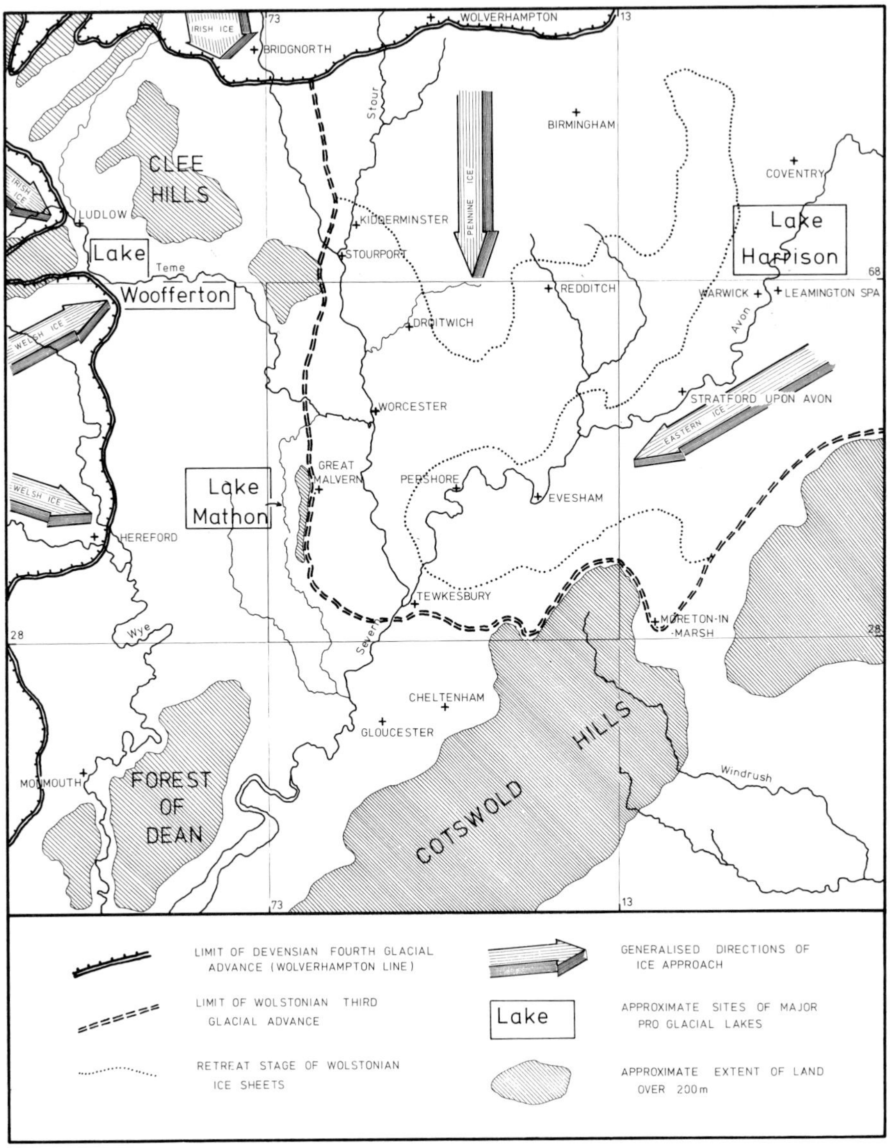

Glacial influences in quaternary times. (BA)

reversal occurred during the Wolstonian glacial advance; the Severn diversions in the early part of the Devensian period, and the Teme reversal between Tenbury Wells and Stockton-on-Teme during the late Devensian, perhaps as recently as only 14,000 years BP.

Quaternary glaciers thus seem to have 'moved' from the area of the Wrekin up to the summits of Plynlimon, the Severn-Wye watershed from the line of the Malvern Axis to the foothills of the Welsh mountains.

The principal evidence for these changes lies in the nature of the river terraces and their constituent gravels. In the case of both the Avon and Teme, accurate field surveys of these terraces show discordant slopes. That is, while the terraces in the valley bottoms show clear slopes in the direction of flow of the present rivers, high level terraces on the valley sides are seen to slope *upstream,* or opposite to the present-day streamflow. Those higher terraces could only have been formed by a river flowing in the direction of the measured slopes. In the valley of the Severn the higher, and therefore older terraces, can be traced upstream only as far as the Ironbridge Gorge, at which point they disappear. Lower (younger) terraces can be traced right through the Gorge and on to the Shropshire plain beyond, marking the event of the extension of the Severn basin. In addition there are significant differences in composition between higher and lower terraces. In some cases this takes the form of pebbles from the far-distant source areas mentioned earlier. For example, in the valley of the Teme, older terraces do not contain materials derived from the Kerry Hills, while the younger ones do.

The dating of much of the Quaternary history of the county has been derived from the application of pollen analysis to a number of peat deposits and by the derivation from organic materials of radio carbon dates. A. Brown and D. Stead have constructed pollen spectra from peat deposits at Callow End, Hartlebury Common and Ipsley Alders Marsh at Redditch. From their diagrams, representing the variation of pollen grains preserved at different depths, it has been possible to reconstruct the vegetational history and through that the climatic fluctuations which caused those vegetational changes. Brown, in 1974, and

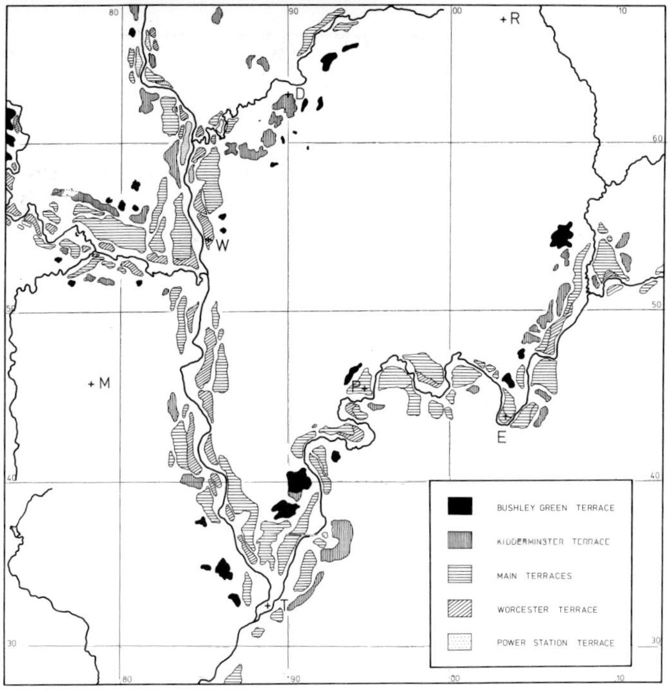

Map showing the river terrace deposits of sand and gravel in the main Severn, Avon, Teme and Salwarpe valleys. (BA)

Glacial period	*Glacial events*	*River terraces*
Flandrian	Rising sea level—infilling of estuaries	Power Station Worcester Avon No. 1 Teme No. 1
Late Devensian	Wye Glacier advance—Lake Woofferton—reversal of part of Teme	Teme No. 2 (lower)
Mid-Devensian	Wills' Wolverhampton Line—Lake Lapworth—extension of Severn. Ironbridge	Main (lower) Avon No. 2 Teme No. 2 (upper)
Early Devensian	no record	
Ipswichian	interglacial	Main (upper) Avon No. 3 Teme No. 3
Wolstonian	ice retreat	Kidderminster Avon No. 4 Teme No. 4
	Eastern ice—Lake Harrison—reversal of Avon	Bushley Green Avon No. 5 Teme No. 5
Pre-Wolstonian		Bushley Green Woolridge

Stead, in 1989, have constructed pollen spectra which show a progressive vegetation change over a period of some 10,000 years, from woodlands dominated by birch in early post-glacial times, to the present-day pattern which is dominated by oaks and elms. A number of sub-stages are recognised when pollen changes depth which indicates variation in the species present. The basal peats, with their plentiful provision of birch pollen and lack of herb pollens, suggest a fairly dense forest growth with a few clearings. As birch reduces in quantity it is progressively replaced by Scots pine, elm and oak, as the climate warmed, culminating in a dominance of hazel around 7,500 BP, the decrease in elm probably indicating early clearing activity by man. This is further supported by the appearance of 'weeds' such as plantain, willow herb and representatives of the family *Compositae* (dandelion-like flowers). Around 2,000 BP the plant community seems to have revolved around a birch-alder-oak mixed woodland. Since then the decline of tree pollens in the counts, together with an increase in herb pollens, supports the idea of much more extensive clearing for cultivation, with oak and alder as dominant trees. More recently the spectra show increases in birch and willows.

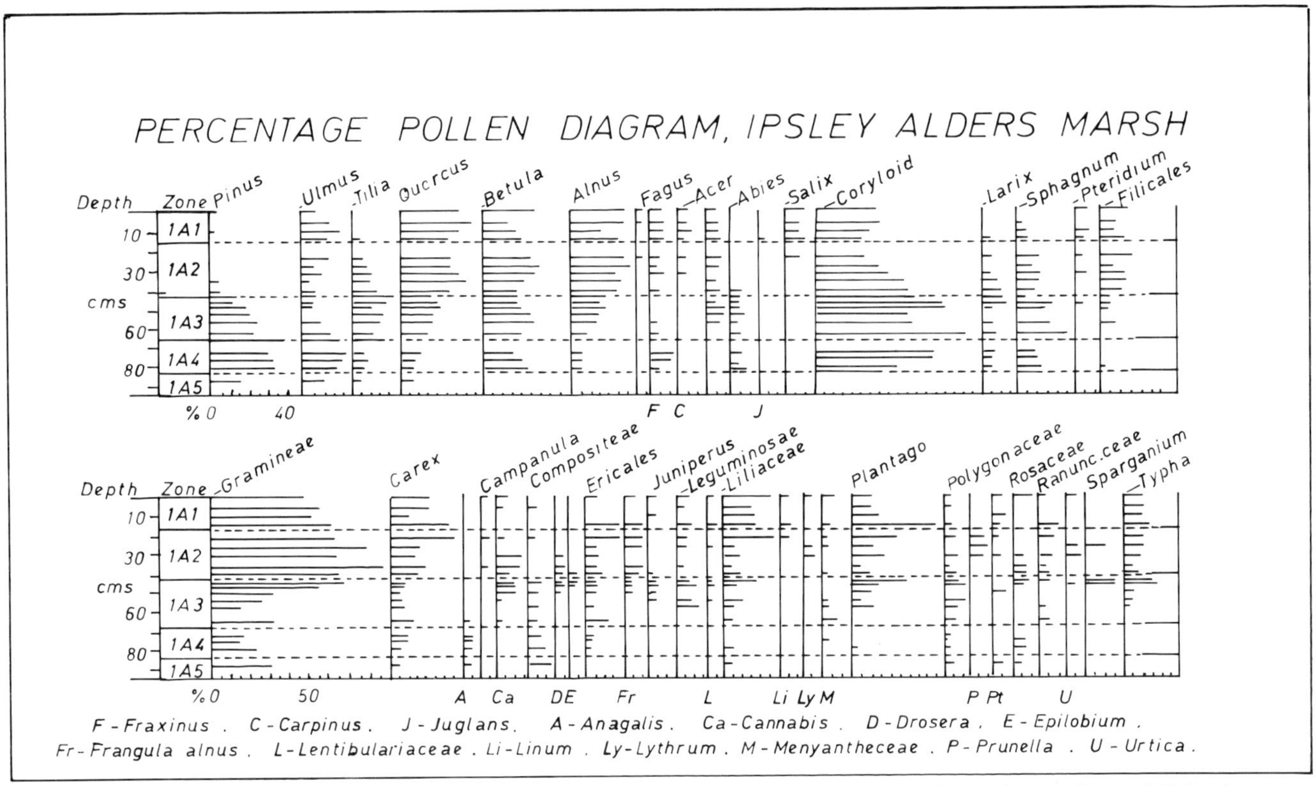

The pattern of vegetation change summarised above corresponds to the well-established pattern of climatic changes since the withdrawal of the icesheets. The immediate post-glacial period is known as pre-Boreal (cold) with successively the Boreal (warmer and drier, between 9,500 and 7,500 BP), the Atlantic (warm oceanic conditions between 7,500 and 4,500 BP), the sub-Boreal (warm and dry again between 4,500 and 2,500 BP), and the current sub-Atlantic period (cooler and more oceanic since 2,500 BP).

Radio carbon dates are rare in Worcestershire, dating therefore being proposed on the basis of those derived from neighbouring counties. From sites such as Cross Mere, Church Stretton, Four Ashes, Woofferton, and Upton Warren, it has been deduced that, while the diversions of the Avon and Severn are of considerable relative antiquity (being older than 38,000 BP) there was glacier ice present in the Ludlow area as recently as about 14,000 years ago and this was the ice responsible for the diversion of the River Teme at Stockton-on-Teme and for the formation of the so-called 'Shelsley Gorge' and 'Knightwick Gap'. It was probably also, indirectly, involved in the shaping of the Longley Green Gap in Suckley parish. Many of the river terraces in the valleys of Severn, Avon and Teme must also post-date this Wye valley glacier, so that the familiar Worcestershire landscape of today is geologically of recent creation.

It is not exactly clear when Man began to have a significant impact on the natural landscape. It has been suggested that Bronze Age woodland clearance commenced as early as 7,500 BP when Neolithic pastoralists created more open grassland for their beasts, and that around 5,000 BP they constructed their magnificent hill forts, such as those of the Malvern Hills, Bredon Hill and the Abberleys. The Roman impact on the county was, in most places, initially military, building encampments on the lowlands and linking them with a web of strategic routeways. In Worcestershire these included the Salt Ways, Ryknield Street and Buckle Street. There followed a period of tranquillity as the Romans extended their occupation and built their rural villas, until the Saxon invasions after the Battle of Dyrham in 577 AD. Then the county was occupied by two principal tribes, the *Hwicce* and the *Weogoran*, the former slowly gaining dominance. This, too, was a period of slowly expanding colonisation, with much more land clearance and settlement, predominantly in the valleys of the Avon and Severn and as far north as the river Salwarpe. North of this and west of the Severn, forests persisted. In contrast to the compact Saxon villages and hamlets set among a patchwork of trim open fields, which characterised the south and east of the county, the north and west contained only a few scattered farmsteads with irregular and scattered, hedged fields. The distribution of place-name endings 'ley', 'leigh', 'heath' and 'field' contrasts quite clearly with that of 'ton' and 'hamton' — clearances in forests as opposed to incipient 'towns'.

While the Saxons were busy preparing the land for cultivation, the advent of Christianity in the region had a contemporary and further aspect, as the Mercian diocese of Lichfield (680 AD) was set up to serve the Hwicce. The initiation of urban centres by the Romans, carried forward by the Saxons, was thus firmly entrenched, and centred on the cathedral city of Worcester, *Weogorna civitate*, in 691 AD. The city has continued to extend its marketing, administrative and ecclesiastical influences ever since. Other 'urban' centres developed around the great Abbeys, such as Evesham and Pershore, and the influence of the church was soon felt further in economic terms as it developed its great estates both socially and economically, being instrumental in much further forest clearance.

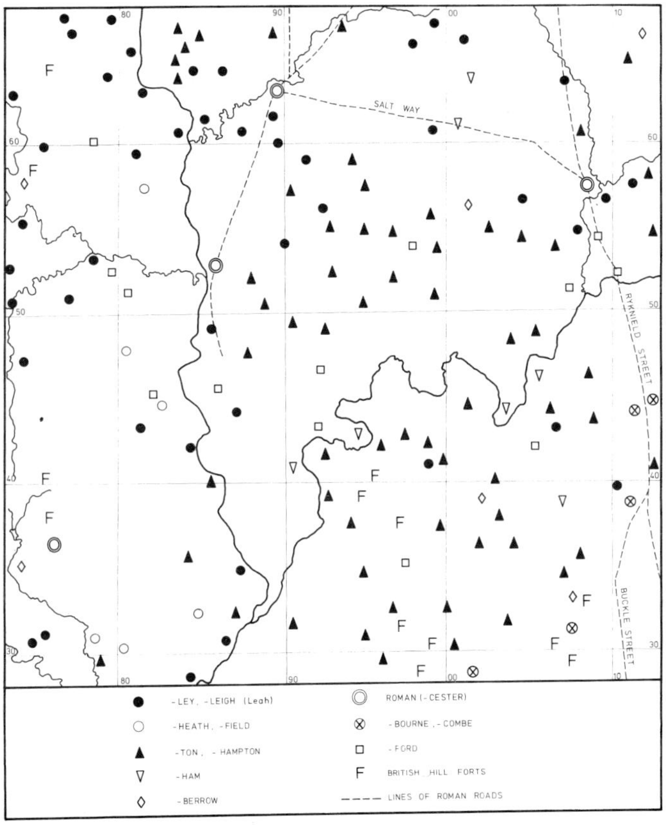

Place name distribution in Worcestershire. (BA)

Danes did not settle in Worcestershire, but the threat of their plundering parties was sufficient to see the establishment of the Anglo-Saxon *burh*, or fortified town, at Worcester around 900 AD. The 'Hundreds' which comprised the territorial 'shire' of the city were largely delimited by the great ecclesiastical estates of the southern part of *Wireceastreshire* (1040 AD). Then came the Normans, after 1066, although it may be said that the 'conquest' of Worcestershire seems to have been accomplished with great ease and little bloodshed if the marked absence of Norman 'mottes' is taken as an indicator. It was the Normans who, pushing the frontier of settlement beyond the Severn and into the Welsh

Marches, were responsible for the final phase of land clearance in the county, within the surviving great forests of Feckenham, Malvern Chase and Wyre Forest. As both ecclesiastical and lay lords granted lands to favoured individuals, there arose moated farmsteads, irregular hedged field patterns, ditch-and-bank boundaries and, in some instances, amalgamations of homesteads to give rise to somewhat amorphous and loosely-knit 'green' hamlets such as Ham Green, Broughton Green and Bradley Green in Feckenham, and at Gilbert's End and Longdon Heath in the Malvern Forest.

The Black Death of 1348 brought an abrupt end to this prosperity and, with far fewer mouths to be fed, agriculture turned instead to the 'lure of the golden fleece', and the rearing of sheep on extensive, enclosed lands. Much of the county's textile, clothing and footwear tradition dates from this period and on to the sixteenth century. The dissolution of the monasteries freed yet more land to private ownership.

The continuing diversification of industry in the county was further increased by mining developments, particularly in the Dudley and Halesowen areas, but so great was the lure and demand for coal that even the low quality deposits around Abberley, Pensax and Mamble were exploited. The industrial development had impact on Worcestershire not in the form of 'Black Country' development, but through the increasing importance of the county's transport system. Initially the roads and rivers were extensively utilised by through traffic of imported raw materials and the export of finished products from Birmingham. As the pressure continued to build, the natural network was complemented, first by the canals of the late eighteenth century, then by the railways. Worcester's importance as a route centre was thus consolidated in the nineteenth century, while the twentieth has seen yet more of the same through the motorway system. The city is still advantageously placed in the national system.

Fortunately, the Industrial Revolution 'passed through' Worcestershire on its transport system, and we are left with a landscape much like that painted so evocatively by Benjamin William Leader RA, founder of the Worcester School of Art in the 1850s, and commemorated by so much of the music of Sir Edward Elgar. That landscape is to this day based on agriculture.

The agricultural variety of the shire is largely due to the quality of its soils. Light and easily worked sandy and silty loams dry out and warm quickly in the spring sunshine, and the climate of the county is mild. The funnel shape of the Bristol Channel and the Severn estuary helps to direct warmer, moist, maritime air from the south-west, right into the heart of the county, at the same time keeping at bay the harshest frosts and snows. An annual average total of around 40 inches (900 mms) of precipitation, almost wholly in the form of rain, ensures the maintenance of good grassland.

Some 90% of the farmed land in Worcestershire is classified as Grade I, II or III quality — an unusually high proportion of 'medium' to 'good' land. Despite many recent profound changes in agricultural practices, grassland is still abundant. Clun, Kerry, Sussex and Welsh halfbreeds are the commonest varieties of sheep to be seen, while the influence of the traditional Herefordshire cow, with its familiar white face, is still readily apparent. Recently, however, farmers have shown much more favour to higher-producing North American bred Friesians, and to an increasing range of European 'exotic' breeds, including Charolais and the Simmental. Fruits and vegetables are traditionally associated with the Vale of

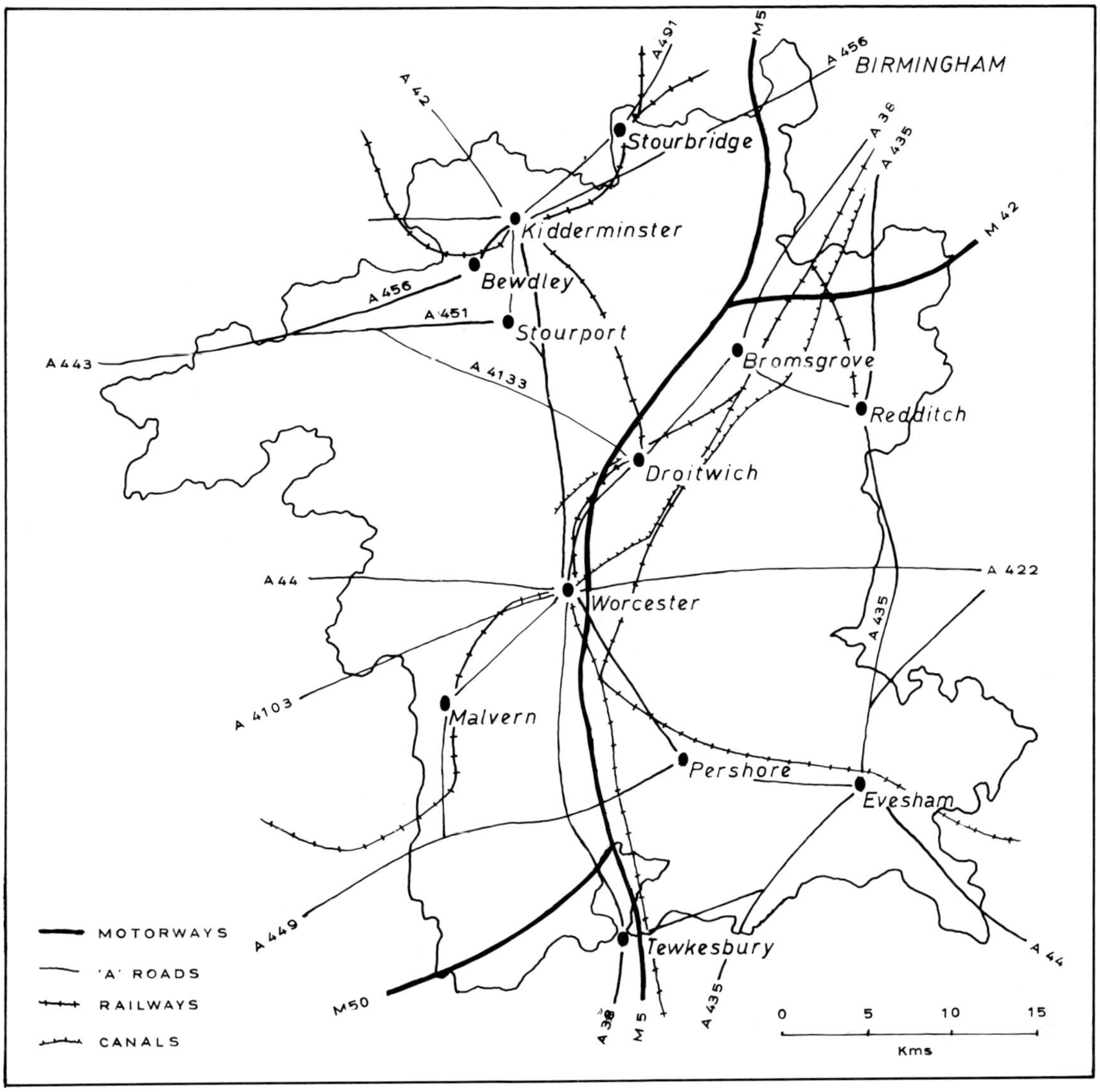

Map of Worcestershire routeways. (BA)

Evesham, where asparagus was once a prolific speciality crop. The Teme valley is almost as well-known for its cherry orchards and hopyards, sad vestiges of which may still be seen by following the springtime 'Blossom route'. During the last three decades, however, intense competition from larger, more specialised farms in other parts of Britain and Europe has brought about a marked decline in traditional crops in Worcestershire, which have been progressively replaced by bread-wheat, barley and, of course, the spectacular yellow of oilseed rape.

ABOVE: The dramatic ancient fortifications instantly identify the Malvern Hills. (DMG) BELOW: Cattle graze the Avon flood plain ham meadows. (DMG)

Severn Vale and Bredon Hill from the Malvern Hills. (DMG)

The River Teme viewed from Osebury Rock. (DMG)

Rivers, Streams and Canals

Worcestershire contains parts of three important rivers — the Severn, Avon and Teme, several smaller rivers like the Stour and Salwarpe and many brooks of different types. Watercourses range from fast-flowing streams in steep-sided valleys to slow, meandering brooks of lowland farmland. There are important canals from the Severn, two leading to Birmingham and beyond.

Virtually all watercourses are of natural history interest, either from the freshwater life they contain or their bankside flora and fauna. Nowadays rivers, brooks and streams and their environs are often important wildlife refuges and corridors in an increasingly damaged countryside. A stream through agricultural land is likely to be the richest habitat for miles around. However, most of Worcestershire's watercourses are polluted, some mildly, like the River Teme, its associated streams, and the Dowles Brook; some badly, like the River Stour. Pollution is of two main types: over-nutrification by inorganic substances (fertilizers) derived from sewage and agricultural fertilizer run-off or by toxic or harmful chemicals like mercury and organo-chlorine pesticide derivatives, which usually arise from industrial activities.

The nutrient levels in watercourses vary naturally depending on the type of underlying geological formations. For example, the clays of south and central Worcestershire provide more nutrients than the sandstones of the Kidderminster district, and the Triassic Keuper beds in the Droitwich area cause natural pollution of the river Salwarpe from upwelling brine: these salt deposits have been exploited since the Iron Age.

Although water chemistry has an important influence on the flora and fauna of rivers and streams, other factors such as substrate, flow rate, light and shade and the amount of suspended particles and the rate of deposition all have an effect. For example, many emergent plants which flourish in lowland slow streams cannot withstand fast-flowing streams over hard rocks. Heavily shaded streams contain few, if any plants but, if the water is pure, there may be many insect larvae feeding on plant debris washed downstream, and fish which eat the insects.

There are several watercourse types in Worcestershire. The main rivers Severn and Avon and many of their tributary streams in central and south Worcestershire are eutrophic (high nutrient level) clay streams. The Stour at Kidderminster is badly polluted and of lesser natural history interest apart from important marshes between the river and the parallel Staffordshire-Worcestershire canal. The Teme is a fast-flowing trout-stream of high-quality water. The Dick Brook is a long, fast-flowing trout stream coursing through wooded valleys and bordered by old wild daffodil meadows.

The Severn is the longest river in Britain, and already a big river when it enters north Worcestershire from Shropshire near Wyre Forest. In Worcestershire, nowadays, it is a river with two faces: above Stourport-on-Severn it has many of the features of a natural, almost upland river; downstream these are lost and the Severn runs broad and deep between fairly high banks. Water levels in the lower reaches are controlled by a system of locks and weirs. In the upper reaches the river is much more varied with deeps and shallows, gravel beds, silty areas, rock shelves and small rapids.

The Severn has been an important transport route since Tudor times. Despite the difficulties of navigation, shallow draught vessels regularly travelled from Bristol to Welshpool in the mid-eighteenth century. These vessels were usually flat-bottomed barges called trows, driven by sails or pulled by gangs of men. Several schemes were devised for building locks and weirs to improve navigation, but the first attempt to obtain a Navigation Act was defeated in Parliament in 1786, because of objections from boatmen who feared loss of livelihood and did not wish to pay tolls, and from landowners who feared flooding. Other schemes were proposed but the only successful early improvement was a towpath for horses, which ran from Shrewsbury to Gloucester by 1812.

Other early work improved navigation on the Stour and Salwarpe by deepening the rivers and building locks. Andrew Yarranton, a remarkable Worcestershire man (1616-1681), did the work, which was only partly successful. Later, the great engineer Brindley cut the Staffordshire and Worcestershire canal between 1766 and 1770. The new cut ran close to the course of the river Stour, from what is now Stourport-on-Severn to Great Haywood in Staffordshire, connecting with the Trent and Mersey Canal. Traffic from the Black Country could then travel to Bristol — when the Severn permitted! Between 1767-71 Brindley cut the Droitwich canal alongside the Salwarpe river, enabling barges from the Severn to reach the saltworks at Droitwich more easily. These canals used the nearby rivers to provide water.

The Worcestershire-Birmingham canal was cut between 1791 and 1815 but, despite these canal connections, the River Severn remained untamed. Nowadays it is quite difficult to imagine what it was like when the depth of water was only 12-16 inches, often dropping to a mere nine inches, and for months boats could only be pulled upstream on a flush of water. After many more years of argument and failed bills, an act was passed in 1842 permitting locks to be built at Lincomb, Holt, Bevere and Diglis. They were open by 1843 and several years of intensive dredging below Diglis were necessary to deepen the river to six feet. Eventually a further lock had to be built at Tewkesbury to provide sufficient depth. By 1858 the Severn at last became safely navigable from Stourport to Gloucester.

The construction of large reservoirs in central Wales changed the Severn, mainly by flood control. Before the days of Lake Vyrnwy, spring and summer freshets were common, and these encouraged upstream migration of fishes such as salmon, trout and lamprey. By 1900 the construction of locks at Gloucester cut direct connection to the sea and stopped the effects of tides reaching Worcester. Flounders, once found at Shrewsbury, soon decreased and there were many other rapid changes in fish populations. For example, the migratory inland flood of elvers (young eels) decreased. Prodigious numbers once swam upstream in spring with an average season's catch of 20-30 million. Estuarine allis shad became rare and the once commoner twaite shad declined.

Despite eutrophication and pollution, the Severn remains a salmon river, and the fish can be seen leaping from the water as they scale Diglis Weir in Worcester on their migration to upstream spawning grounds. Apparently small numbers of flounders still reach

Worcester, for one was found during dredging operations in the Laughern Brook in the 1980s. Nevertheless the Severn is primarily a river for coarse fish like perch and roach together with the predatory pike, although introduced barbel are common nowadays.

Barbel are introduced fish which have spread widely throughout the Severn. (RMB)

Competition from the railways eventually decreased usage of canals and waterways: the Droitwich canal was abandoned in 1939. During World War II improvements at Diglis Dock in Worcester increased river use, particularly for transporting petrol to Worcester and Stourport during the 1950s. Since then commercial traffic on the Severn has virtually ceased. The upper reaches above Stourport remained unimproved for navigation and traffic gradually declined in the late 1800s.

Nowadays, a great resurgence of interest in waterways for leisure activities, especially boating and angling, has led to the re-opening of many canals. During the last few years voluntary labour and the Government's job creation schemes have been slowly clearing the Droitwich canal, with the resultant destruction of the county's largest colony of reed warblers. Between the closure in 1939 and the 1980s, the canal developed a bed of common reeds about five miles long, colonised by many birds, especially reed warblers. No actual estimates of numbers are available, although singing birds were found every few yards in many places and hundreds were caught for ringing. Some ringed birds moved between the canal and nearby Westwood Great Pool.

During the last ten years new proposals have been made to open the upper Severn for navigation by leisure craft, building locks and weirs, and dredging or blasting to make a navigable channel. Many people are opposed to this for fear of disturbance and pollution and, most important, because of the imposition of uniformity on a wild stretch of natural river, so decreasing the diversity of habitats and species.

The history of Severn navigation is important in considering the natural history of the river. A shallow, clear, varied river contains a different and greater assemblage of animals and plants than a deep, turbid, canal-like one. The upper Severn in Worcestershire perhaps gives some idea of the nature of the ancient unspoiled river and we are extremely fortunate that the stretch north of Bewdley remains little changed. It is picturesque and attractive as it winds gently through the wooded valley between Wyre Forest and Eymore Wood from Upper Arley. In summer the great drifts of water crowfoots in the shallows are starred with white flowers. These plants are excellent cover for freshwater invertebrates, and dragonflies abound. Banded agrions are often prominent in courtship display, perching on these weeds among the swirling water. White-legged damzelflies are common here, where pollution is minimal. Beneath pebbles and stones there are stonefly and mayfly larvae and a sweep by net through shallow, fringing vegetation reveals abundant waterboatmen and water beetles. A spectacular denzien of shady banks is the giant lacewing with glowing orange eyes and clouded wings.

A rare dragonfly of this area is the club-tailed dragonfly. Because of its elusive behaviour it is often overlooked despite its large size (average wingspan of 64 mms) and dramatic appearance. It is nationally uncommon, but recent records suggest the Severn in south Shropshire and north Worcestershire is an important stronghold. The adult insect is on the wing for a relatively short period of about three weeks between end of May and early July. Synchronous hatching is usual and the insects quickly disperse away from the river, often into nearby woodland and scrub. After the juvenile (teneral) period in which the colours become fully developed, the males hold territory over calm water near trees while the females remain perched on trees and shrubs. After mating she lays eggs in shallow stretches of water. The young larvae burrow into soft, silty substrate and, when they eventually mature, they climb out of the water on vegetation or other objects and the adults emerge. The club-tailed dragonfly is most numerous in the natural river above Stourport, and rapidly disappears further south although, in the last few years, a few have been found elsewhere in the county. They are insects of shallow waters in big rivers.

Water birds are common. Mallards nest where there is enough cover near the river — usually on the ground beneath vegetation. Pied wagtails nest by the river, on a building or ledge, and forage for insect food on open ground along the water's edge. The less common grey wagtails are usually associated with rougher waters and rocky areas, near bridges and mills. They nest on the ground or on ledges, and feed largely on insects found on the ground at the water's edges or caught in the air. Cormorants dive and hunt for fish on the river and at nearby Trimpley reservoir outside the breeding season. Small numbers of mandarins live in Wyre Forest and occasionally consort with mallards on the river. Even goosanders turn up regularly, possibly reflecting the increase in Welsh breeding birds.

South of Bewdley to Stourport the river becomes more uniform, although there are shallows which small boats can pass with care when there is plenty of water in the river. South of Stourport uniformity reigns. The water is deep and dark — too dark and deep for water plants to grow other than near the bank, which rises sharply from the water for six to ten feet, occasionally providing nest sites for kingfishers. A narrow fringe of emergent plants such as bur-reed and purple loosestrife provides shelter for aquatic insects. The banks are subject to considerable erosion, especially in times of flood, and areas of bare soil are colonised for a while by arable 'weeds', such as black mustard. Tree and shrub cover is rarely as great as along the upper Severn: willow, alder and hawthorn scrub are common and a narrow broken bank of scrub is a feature. In winter flocks of siskins and redpolls extract seed from alder false-cones. The river is much more open in the lower reaches and bordered by large floodplain meadows, especially south of Worcester. In a few places there are associated marshlands. For example, north of Worcester several wetlands, grouped into an SSSI, butt onto the river, some in old pits excavated for clay.

The big river attracts migratory birds which appear to follow its course. A day on Worcester bridge bird-watching, during migration times, can be surprisingly productive. Parties of common sandpipers may flash through and cormorants are common. Gulls are seen throughout the year, less so in summer, although even lesser black-backed gulls now occur there as small numbers breed on rooftops in Worcester. Gulls have become more common inland during the last fifty years, probably because of the great abundance of food at rubbish tips. In winter the most common is the smaller black-headed gull, often seen along the river in Worcester, where it scavenges. This species, lesser black-backs and herring gulls gather in large numbers at municipal refuse dumps. Flocks of hundreds of gulls often drift over Worcester. The larger gulls roost mainly on the Severn estuary at night and move inland during the day to forage. In the evening they can be seen high up, flying south again in an arrow formation, although some gulls roost at Westwood Great Pool.

The river Severn floods during most winters and water spills over miles of riverside meadows. In Worcester the Pitchcroft racecourse soon becomes a large lake, much to the delight of the gulls, which bathe in the water and perch on the racecourse rails to preen and rest. Worcestershire's famous County Cricket ground, in the shadow of the riverside cathedral, is also often flooded. Pastureland floods force invertebrates (earthworms and

OPPOSITE ABOVE: Clay stream with abundant aquatic and emergent vegetation, and also with sprouting pollarded willows. (GHG) BELOW: Sedge warbler feeding young. (ARH) ABOVE RIGHT: Giant bellflower, a plant of western brooksides CENTRE: Yellow loosestrife. (both DH) and BELOW: greater dodder, a speciality of the lower Avon. (GHG)

insect larvae) to the surface or they are drowned and drift to the flood margin. These are rich pickings for birds and big flocks soon gather. Species include large flocks of wintering lapwings, starlings and gulls. Carrion crows, jackdaws, rooks and pied wagtails all join in.

Mute swans are a feature of the Severn at Worcester with flocks eager for food regularly gathering near South Quay. Numbers have varied greatly over the years with flocks of up to fifty birds in the 1960s. Numbers then rapidly declined in parallel with similar dramatic decreases in many parts of southern England. Research showed this to be due to lead poisoning caused by eating discarded or lost split lead shot used by anglers as line weights. The shot is presumably mistakenly eaten as grit or seeds. In the late 1980s the use of lead was banned, but river banks are still contaminated. In recent years the Swan Study Group has become active in the county and, with the help of sympathetic veterinary surgeons, has caught poisoned swans for treatment, to return them to the wild. Swans injured by entanglement with fishing line, swallowing hooks and colliding with overhead cables have also been nursed back to health.

Three national surveys of mute swans have been organised in the last twelve years by the British Trust for Ornithology and the Wildfowl and Wetlands Trust. Total numbers in Worcestershire were 46 in 1978, 49 in 1983 and 272 in 1990. Hopefully these figures reflect a decrease in lead poisoning though, of the 272 birds found in 1990, 165 had been treated by the local swan hospital, the majority for lead poisoning. In part, the recovery of numbers in Worcestershire reflects the dedicated work of a small band of people. Sadly, angling still presents a threat to swans and other animals and birds. Large quantities of discarded line, complete with hooks and sinkers, are still found on river banks — nylon line does not rot — and many birds have been found severely tangled or pierced by hooks.

Shakespeare's Avon rises in Northamptonshire at Naseby, traverses Warwickshire and south-east Worcestershire, to join the Severn at Tewkesbury in Gloucestershire. It enters Worcestershire from Warwickshire in a rather shifty fashion, as the county boundary runs down the centre of the river for about a mile at Cleeve Prior. An excellent view of this area is obtained from the top of the bank of the WNCT nature reserve at Cleeve Prior, south of the river. Here the river flows close to a steep, wooded area in the reserve, with extensive flood plain meadows on the far bank, partly covered by a caravan park.

In Worcestershire the Avon follows great meanders through a wide flood plain and passes through Evesham and Pershore. Although Cleeve Prior is only 15 miles from Tewkesbury as the crow flies, the river winds for 31 miles round the north and west flanks of Bredon Hill.

Unlike the Severn, the Avon has a long history as a navigable waterway. William Sandys of Fladbury first conceived the idea of a navigable Avon, and he obtained an Order in Council from Charles I for establishing a Royal Commission to assist. Operations commenced in 1636 and the navigation was in use, but not complete, by 1640. Nine locks or watergates were built in Worcestershire — Strensham, Nafford, Pershore, Wyre, Fladbury, Chadbury, Evesham, Harvington and Cleeve — usually downstream of a mill. Some of the 'locks' were in fact 'watergates' which narrowed the river and restricted flow, so keeping water levels higher. Boats had to be worked upstream against the flow through the gate and often had to wait for long periods for the next section of river to fill, after the gate had been closed.

The Avon in Worcestershire is a fairly broad, slow-flowing river, falling only about 11 metres over 31 miles. Although it has a winding, natural course, it is in many respects a man-made river, repeatedly dredged during three centuries. It contains many locks and weirs and is heavily laden with nutrients from the outflows from many sewage works, and from agricultural fertilizer run-off. In recent times the Avon has suffered pollution by toxic chemicals, probably originating from Coventry. Particular concern has been caused by high levels of organo-chlorine insecticide residues found in eels and dead herons.

Water crowfoot forms great beds of flowers in the Upper Severn. (LS)

The types and levels of organic nutrients found in silts and water greatly influence the plants found in a river. The higher the nutrient level (the more eutrophicated the river) the lower the species diversity. Some species do well in eutrophic rivers, some are tolerant, others intolerant. The Avon runs over Lower Lias clays which naturally contain a higher and wider range of nutrients than, for instance, hard insoluble rocks and sandstones. This and further eutrophication from sewage limits the numbers of plant species to those able to tolerate such conditions, although in the Avon there is often some recovery of species diversity between sewage outflows.

Other factors also influence river plant life. For instance, the Avon is a deeply dredged, turbid river with soft mud and silts frequently disturbed by the passage of boats. Little light, therefore, reaches the depths of the river and plants cannot grow mid-stream. Dredging has usually left steep banks which are difficult for plants to colonise, although this structure is frequently modified by erosion, which produces a narrow shelf at the base of the bank covered with shallow water on which emergent water plants can grow. Among established plants and on the inside of bends where water flow is slowed, silt deposition increases and mud banks form on which more plants grow, and the fringe of emergent plants becomes wider and more varied.

The Avon is classified as an eutrophic clay stream and its typical and most frequent emergent plants are yellow water lily, true bullrush, great yellowcress, bur-reed and arrowhead, species which also tolerate sewage-enhanced eutrophic conditions. Common reed forms long narrow beds in water with a maximum depth of about one metre. Other less common emergents, indicating less eutrophication or cleaner habitat, are duckweed, great water plantain, reed grass, and flowering rush. Reed canary grass grows on alluvial banks.

Reed sweet grass grows in shallow water at margins or in swamp conditions. Patches of reed mace are frequent but lesser reed mace is uncommon. Amphibious bistort is common away from sewage outflows, growing in wet soils and often spreading out from the water's edge into deeper water. Its oval, flat leaves float on the water surface and clustered pink flower spikes are thrust a few inches into the air. The charmingly-named galingale adds spice for plant hunters here and there.

Although this array of plants is relatively small it is attractive. It does not approach the species richness of chalk streams, but it gives an appearance of luxuriant growth along the river bank, and quickly re-grows after dredging, if bank structure allows. The plant species tend to grow in discrete clumps so there is an appearance of greater variety than is actually present. As silting

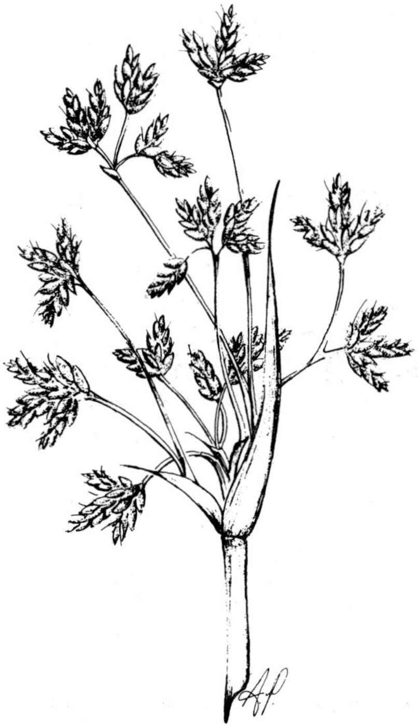
Bullrushes are common along the Avon. (WNCT)

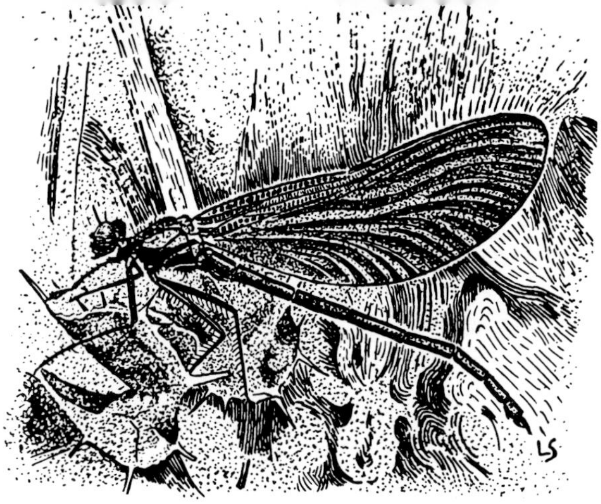

Agrion virgo, a scarce damzelfly of fast-flowing streams. (LS)

increases, and the margin of shallow water and mud widens, various other wetland herbs appear. These include water forget-me-not, water mint, fool's watercress, water speedwell, brooklime, great pond sedge and water chickweed. These wider fringes of vegetation resemble ponds and often have similar muddy, trampled edges where cattle drink or stretch out to graze their favourite sweet reed grass. In these disturbed areas celery-leaved buttercup and trifid bur marigold are common.

The shallow water among the plants contains many freshwater insects and their larvae. Common dragonfly larvae are southern aeshna, brown aeshna, banded agrion, white-legged damzelfly and the common blue damzelfly. Banded agrions are a striking feature of the Avon valley as the adults are often common in early summer. In recent years there have been a few records of club-tailed dragonflies to add to those from the upper Severn. The nationally rare white-legged damzelfly is abundant on the Avon as well as along the upper Severn, making Worcestershire outstandingly important for the species. Aquatic predatory bugs include the oddly shaped water scorpion with its large clasping front legs, waterboatmen and pond skaters.

The Avon is bordered by steep banks of variable height, depending on nearness to lock and weir. In places the bank rises steeply for ten feet or so; in others the water laps to within two feet of the bank top. The banks are often bordered by large flood plain meadows, grazed by cattle and sheep and cut for hay or silage. The banks become eroded in places and colonised by arable weeds. In places the bank is backed by a low ridge originating from dredgings, and behind this there are sometimes low-lying areas of wet field or marsh. This in turn may be backed by a river terrace of gravelly, well-drained soil — excellent market-gardening land. This type of structure is particularly obvious in the Pershore-Defford area and, left to itself, it supports a wide range of damp or wet habitats. The river bank behind the emergent plants is often clothed with gipsywort, marsh woundwort and clumps of skullcap.

Marsh woundwort and nettles in places along the Avon between Eckington and Pershore have in recent years been covered with wiry reddish coils of great dodder. This unusual plant, which is rare and apparently decreasing in Britain, is a parasite (like all the dodders) and although it is mentioned in old accounts of the county flora there were no recent records until many plants were found in 1988. On germination dodder puts up a small shoot which must soon contact a host plant or the seedling dies. If contact is made dodder grows, climbs and coils round the host, attaching itself with haustoria or 'suckers', through which food is obtained. The dodder seedling's root shrivels and the plant, which does not contain chlorophyll, becomes entirely dependent on the host. Clusters of pinkish flowers form close to the stems, though the plant does not produce leaves. The seeds of dodders are known to survive for long periods and it may be that riverbank maintenance disturbed buried seed causing the sudden appearance of 1988. A similar plant, flax dodder, used to occur in Worcestershire when flax was a common crop in the eighteenth and early nineteenth centuries. Lesser dodder occurs as a parasite of gorse and one variety was once a serious parasite of clover crops.

In drier areas behind the river bank there are large beds of stinging nettles and in places many brassicas like black mustard, whose seed crop is rapidly eaten by linnets and greenfinches. In places there are great stands of the six-feet tall, poisonous umbelliferous plant hemlock with delicate, lacey leaves and red-spotted stems, smelling of mouse. In the damper areas marshes develop, dominated by great hairy willow herb, yellow flag and meadow sweet. Where the water table is at the surface large patches of reed sweet grass, great pond sedge and more uncommon plants such as tubular water dropwort grow.

These areas of wetland and nettle bed are rich in insect life and excellent breeding and nesting grounds for many bird species. Nettles support over 100 species of insect and spiders, harvestmen, woodlice and snails are also common. There is a complex web of life in a nettle bed: many insects are sap-suckers, including abundant aphids and plant bugs, and they all have their predators, which are often beetles, including several species of ladybirds. Other predators include lacewings, hoverflies and earwigs. When cow pastures are near, a common predator is the yellow dung fly which can often be seen carrying another species of fly which it has caught to eat.

Peacock, small tortoiseshell, red admiral and comma butterfly caterpillars all feed on nettles. The colonies of black small and yellow, spiney tortoiseshell caterpillars on a silken web are perhaps the most familiar. The comma caterpillar, also found on elm and hop (wild hop is quite often found climbing nearby hedges), is effectively marked to look like a bird dropping. Commas are, unusually these days, a butterfly success story. At the beginning of the century they were restricted to Herefordshire, Worcestershire and Monmouthshire but now they are widely distributed.

The number of woody and bushy plants along the river bank is variable. In places there are none; in others there is a scattering of hawthorn, blackberry and dog rose, probably growing from seeds brought in by birds, and in others a continuous band of scrub. Osier willows are common and English alder grows in some places and may form big trees. Willows are a common feature of the Avon valley as scrub, trees or pollards. Willows produce abundant wind-blown seed and wet open areas are quickly colonised when the opportunity arises, such as on the mud beds of the settling tanks near the waterworks at Strensham, where dense young willow scrub can grow six feet in a couple of years. Pollarded crack willows or white willows are abundant in the Avon valley, growing in rows on the river bank, in hedges, near ponds and in open fields. Many pollards were deliberately planted long ago by thrusting poles into wet ground. Willows are easy to grow and soon produce a crop of poles from the top, out of reach of grazing animals — a constant supply of temporary fencing timber, for willow soon rots. Cut every 10-20 years, the tree quickly grows again and again perhaps over 100 years. As the tree gets older, the heartwood rots, leaving a living external shell and crown, soon house and home to many other plants and animals.

The rough crowns collect rotting leaves round the bases of the branches and a compost forms in which seeds brought in by birds can germinate. Gooseberry, elder, hawthorn, currant, blackberries and dog rose are all common pollard willow plants. Occasionally wind-blown seeds of other trees arrive, germinate and send roots down through the dead heartwood of the willow, forming intricate composite trees. Mixes of ash, hazel, elm and willow are sometimes seen.

Old pollarded willows provide good nesting sites. Mallards often nest in the crowns and wood pigeons, blackbirds, song thrushes, chaffinches and many other common song birds use willows as nest sites. Wrens often tuck their domed dead-leaf nests into cracks in the trunk and holes and hollows are used by jackdaws, little owls, stock doves and tits. Redstarts once nested in pollarded willows in the Avon valley but disappeared with the national decline of the species in the 1960s and early 1970s, due to drought conditions in the bird's wintering quarters in the African Sahel region. Although many British breeding populations have now increased to former levels, those by the Avon and on farmland areas have not done so.

The true osier willow is often seen along the river bank, especially where osier beds used to grow. Osiers have long been a feature of the district: 'Pershore' may well derive from Old English words meaning 'osier-bank'. A large supply of big, flat baskets (skips in local parlance) were required for horticultural produce and osier beds were once large and common in the Avon valley. Many small areas and several large fields, which are now pastureland or arable or have been cleared and drained, are marked as osier beds on maps c1900. Osiers were often grown on short pollards a foot or two high and cut every year or two. Several species of willow were used to provide 'rods' of different colours. Only a few remnants of osier bed now remain, one on the lock islands at Strensham. In Gloucestershire and along the Severn osier beds were often planted in pits excavated to provide clay for brick and tile making. Apparently rods of willow were thrust into the pit behind the diggers, to provide them with the best possible start in life.

Sadly many agricultural and river engineering 'improvements' in the Avon valley in the last 15 years have reduced the diversity and richness of its natural history. Many small wet areas, disused osier beds and scrubby areas have been cleared and most riverside fields have been drained. Brooks and ditches have been deepened and widened. The Avon itself is regularly dredged. Leisure activities also cause river edge trampling and disturbance. Fishing and boating are popular.

An idea of previous richness can be gained from a census of breeding birds in 70 acres of mixed habitat near the confluence of the Avon and the Bow Brook at Defford. In 1971 434 pairs of birds of 42 species, including 22 pairs of marsh warblers were recorded in this area. The richness and abundance was due to the mosaic of wetland, scrubland, old trees, hedges and open spaces. Since then the Bow Brook has been deepened and widened with a summer water level now much lower. The water table in riparian wetlands now dips sharply to the brook water level and areas which previously remained wet all summer are now drained. Many shrubs and trees have also been removed and adjacent fields have been drained and cultivated. This pattern of change has spread through the Avon valley, simplifying habitats and removing many rough and damp corners.

The Worcestershire Avon valley is famous for its marsh warblers. This summer visitor has fascinated bird watchers for many years because of its rarity, odd national distribution and remarkable song. The species itself was one of the last European birds to be described in the scientific literature because of its resemblance to the reed warbler. Savi made the first descriptions in 1824; the bird was recorded Britain in about 1840, and Ward Fowler put an end to doubts about English records when he found marsh warblers near Oxford in 1888. Tomes, writing in the *Victoria County History* in 1901, claimed that marsh warblers had bred in Worcestershire in the Avon valley for 30 years prior and reported that he had shot birds at Littleton in 1887 to confirm identification.

Marsh warblers are elusive and often pass unnoticed. They are the last summer migrant to arrive in England in spring. Males arrive before the females in the last week of May or the first week of June and immediately adopt and defend a territory.

They sing vigorously and continuously until they attract a mate. After pairing, singing becomes intermittent and, although the birds are seen nest-building, as soon as the female starts to incubate the birds stay well-hidden. The territory is usually of tall, rank, marshland herbaceous vegetation such as great hairy willow herb, meadow sweet or nettles, near running water with some trees or shrubs. Sometimes the birds nest in drier situations among tall arable weeds and in 1938 Harthan reported seven pairs nesting in a bean field 200 feet above the Avon. The nest is set in the herbacious vegetation, attached to several stems with 'basket handles' of dried grasses and other materials. The clutch of 4-5 eggs hatches in about 12 days and the young are then fed on insects collected by both parent birds. The richness of the habitat becomes apparent when adults are watched feeding chicks — they collect insect food quickly and easily within a few yards of the nest. The young fledge in about ten days and remain in the area for a week or two. The adults and young migrate almost immediately after nesting and have often departed by the end of July.

How many marsh warblers breed in Worcestershire? In 1990 there were five pairs, but earlier in the century there were many more. In 1912 Eliott Howard, a Worcestershire business man and brilliant ornithologist, wrote in *British Warblers* of marsh warblers 'more plentiful than might be thought' extending their range north of Worcester 'by no means uncommon especially in Gloucestershire and Worcestershire . . . the Severn valley is especially favoured . . . how rare the bird was supposed until we looked for it'. His experience was similar to Harthan's in 1938 '. . . local but not uncommon in the lower Avon Valley where they nest in osier beds and similar moist places' and in 1947 when he found them in many places in the Teme, Avon and Severn valleys. However, in the 1960s marsh warblers could not be found along the Teme and were scarce along the Severn, although we found about 100 pairs in the Avon valley. Since then the species has declined to a solitary male in 1989 and, much to our surprise, risen to five pairs in 1990.

The reasons for the decline have been subject to much research and surmise. Research by Martin Kelsey in the early 1980s concluded that habitat loss and climatic trends in the 1960s and 1970s gradually pushed the isolated Worcestershire population down to such a low level that it became vulnerable to random factors such as bad weather for a season or losses on migration, and demographic factors such as the difficulty in finding a mate when the population became small.

Worcestershire had been the stronghold for British breeding marsh warblers for decades, with a scattering of birds throughout southern England. The Worcestershire birds appear to have been isolated with no evidence of emigration from elsewhere in England or from continental populations. Will marsh warblers return to Worcestershire? Perhaps. There is some evidence of an influx of breeding birds into Kent in recent years and continental birds may colonise England — the species has been extending its range in Europe. WNCT has established many small reserves for marsh warblers: 60% of the British population bred on WNCT reserves in the early 1970s. These reserves still remain fairly intact despite increased drainage, and await re-colonisation. They are now also important areas of marshland, containing many interesting plants, birds and animals.

Water voles used to be the commonest mammal along the Avon. The sound of plops and splashes of the rat-sized vole, about 200 mms long, leaping for safety into the water when disturbed, was a common experience in the 1960s. They are agile swimmers which build grassy ball-like nests in holes in river and stream banks, and they are vegetarians. Water voles are now scarce and their decline is often attributed to the increase in mink, common in Worcestershire, but the connection is not proven. Similarly, several riverbank birds, notably moorhens and mallards, have also decreased. Studies of mink have shown that they are highly predatory carnivores which do eat all these species, but they eat far more

Water vole, once common but now scarce along waterways. (BW)

rabbits, and there are examples of sites where mink are common and the same species of birds and water voles have not decreased. The disappearance of water voles may in part be due to mink and part due to increased disturbance, or river bank and brookside engineering.

Upstream of Knightsford bridge (where the Teme turns east and cuts through the West Worcestershire Hills) the Teme valley, with its steep sides and dingle valleys, woodlands and fast-flowing river is different from the rest of Worcestershire. Downstream the river meanders rather more widely over the red marl plain to join the Severn near Powick. The Teme cut its valley during the last Ice Age, about 14,000 BP, after diversion from the Wye system. Huge volumes of water carved out a deep valley. Later the present river, carrying much less water, cut its own channel in the floor of the original valley, often changing course to leave ox-bows and backwaters which may be seen as lakes or scars on the valley bottom. The river runs in a deep channel and used to cause summer flooding of adjacent farmland, now less frequently, due to channel and bank clearance.

Many present-day short, tributary streams rise at a spring line about 150-200 metres above sea level high up on the valley sides, especially on the south and west sides of the valley above Knightsford. They are fast-flowing, plunging to the river about 150 metres below and actively cutting deep precipitate valleys or gorges, known as dingles, in the valley sides. The streams are cutting through the beds of the red marls and sandstones of the Ditton and Downton series of the Old Red Sandstone. Between the two on the south and west side of the valley there is often a narrow layer of Psammosteus Limestones. Many springs rise from this and, where the streams flow over masses of mosses and vegetation debris, they deposit carbonates, slowly turning the whole mass to a soft rock called 'tufa' or 'travertine'. Southstone Rock is one of the biggest deposits of tufa in Britain, and it was mined in the past and used to build local churches, notably at Shelsley Walsh and Eastham. Most of the dingles are near Stanford-on-Teme and Shelsley Walsh and rejoice in names like Witchery Hole, Devil's Den, Temple Dingle and Death's Dingle. The streams are good for mayflies and dippers and the damp rocks are often clothed with abundant mosses, liverworts and ferns. They run through old woods.

The Teme floods fairly often in its lower reaches, between Broadwas, Bransford and the Severn, usually because its outflow is slowed and backed-up by flood waters of the Severn. Otherwise the deep channel of the Teme is usually adequate to hold the heaviest spates arising from its own catchment.

In 1979 J. J. Day carried out a detailed survey of 57 km of the river Teme, from its confluence with the Severn to Little Hereford above Tenbury Wells, for the WNCT, and much of the following is based on his observations.

The Teme is outstanding for its diversity and 'naturalness'. The banks are varied in height and steepness and consist of both soil and rock; the river bed contains a patchy mixture of silts, sands, gravels, coarse stones and bed rock, and these may be uncovered and dry or damp in periods of dry weather; water flow rates vary from fast to mill-pond-like areas. The soils also vary in type and may be acidic or limey. Management of the banksides is varied, ranging from arable land to grassland, rank vegetation, scrub and woodland. As a consequence, the river, its banks and the valley slopes support a wide range of flora and fauna.

The Teme is a good fishery. The dominant species in the slower-flowing parts are chub, dace and gudgeon, with an influx of introduced barbel in recent years. Roach are fairly common. Eels are the main predatory fish and pike and perch are widespread. In fast-flowing areas, brown trout are fairly common and minnows, bullheads, stone loach and brook lamprey occur widely. The Teme has long been renowned as a grayling river above Knightsford, although numbers declined severely after the 1976 drought — these fish require cool, highly oxygenated water and the temperatures reached during the drought were beyond their tolerance. There has been some recovery and the fish remained frequent above Tenbury Wells. Six species of migratory fish have been known to breed in the Teme: salmon is the most prominent and the number of rod-caught fish in the river may reach nearly 20% of those caught in the Severn catchment. Salmon spawn in suitable gravel shoals. Sea trout occur widely in small numbers; river and sea lampreys spawn near Bransford and in good years twaite shad and allis shad may be found. The lowest two miles or so of the Teme, before it joins the Severn, have similar characteristics and the coarse fish of that river.

Mink are now common, having increased greatly in the last twenty years. As on the Avon there has been a simultaneous decline in once-common water voles. The mink is the dominant predator. The Teme is important for otters — once common. Despite nation-wide declines and the almost complete disappearance of otters from Worcestershire, a few did hang on along the Teme, especially in the higher reaches beyond Worcestershire. There are still many areas of undisturbed suitable otter habitat, such as wooded islands, and in recent years there have been signs of a recovery with an increasing number of records. There is hope that the Teme will be the reservoir from which otters spread back into the Severn and Avon valleys.

As might be expected from the excellent fish population, the Teme is a good river for invertebrates. These have been partly investigated by the Severn Trent Water Authority during monitoring for pollution, and by other entomologists. Eighteen species of stonefly have been recorded, including several national rarities. Twenty species of mayfly, 28 species of caddis fly and many water beetles, one probably new for Britain, have also been found.

The aquatic and emergent communities of plants are fairly restricted, because of the structure of the river and the rather harsh conditions, with few bank-side slow-flowing shallows or mud banks. Most frequent species are long-leaved water crowfoot, spiked water-milfoil, amphibious bistort, reed grass, great yellow cress and bur-reed. Less common are fennel pondweed, perfoliate pondweed, common water crowfoot, flowering rush and lesser duckweed, and there are a few uncommon species. The river margin has a richer flora where conditions are less harsh and typical plants are water chickweed, brooklime, purple loosestrife, monkey flower (an introduced plant which is now well-established throughout Britain), yellow loosestrife, hemlock water dropwort, water forget-me-not, fool's watercress, water pepper, wood club-rush, woody nightshade, marsh woundwort, great hairy willow herb, corn mint, creeping yellow cress, green watercress and celery-leaved buttercup.

Bank vegetation ranges from almost bare ground to closed canopy woodland. Bank-edge woods are scarce and, where they occur, the common species are crack willow, white willow and alder, frequently mixed with ash, oak and poplar. There is an abundance of woody scrub of varying heights and ages, including areas of new colonisation of shingle beaches, and others of dense thickets growing from coppiced bankside trees, often following management work by the Water Authority. In areas of colonisation osiers tend to be the first invaders, followed by crack willow, white willow and grey willow, and then alder. The ground flora varies with the density of the canopy, the amount of shade and nearness to the river. For example the ground flora in osier thickets on shingle beaches tends to be dominated by reed canary grass. In the more mature willow-alder scrubland there are close similarities to older woodland. Common species are cow parsley, herb Robert, ground ivy, garlic mustard, rough-stalked meadow grass, slender false-brome, white dead nettle, nipplewort, red campion, wood bitter-cress, hedge woundwort, and ground elder. Open willow-alder communities contain species of the tall herbaceous bankside vegetation, which is varied depending on shade, wetness and grazing pressures. There are extensive nettle beds and dense stands of

Kingfishers are a familiar bird of rivers with steep banks, in which nest-holes are excavated. (RMB)

butterbur, (the latter with early tall spikes of flowers in spring and enormous leaves in summer); plants of damp conditions like great hairy willow herb, wild angelica, hemlock water dropwort and meadow sweet. Other common species include hemlock, mugwort, false-oat grass, bindweed, rosebay willow herb, tufted hair grass, perforate St John's wort and wild cabbage. All these riverside plant communities provide cover for a wide range of birds, mammals and invertebrates.

The vigour of the Teme often creates new open ground conditions by bank erosion and associated small landslips, and by deposition. Kingfishers and sand martins excavate nest-holes in the Marl cliffs common in the lower 12 miles or so of the river. The plants of the earthbanks include daisy, marsh cudweed, figworts, mullein, horsetails and weld (a plant once cultivated for the yellow dye prepared from its leaves). Green figwort, a nationally rare plant, is moderately common along the banks of the Teme, though it now faces competition from the inevitable Himalayan balsam. In places the small landslips have carried small pieces of old unimproved grassland away from agricultural treatment and these remnants contain plants like lady's bedstraw and ox-eye daisy.

There is an active movement and deposition of gravel within the river channel and all stages of development can be seen from the formation of sand and gravel bars to fully wooded islands, and areas formed in the past and claimed for agriculture.

The birds of the Teme are typical of a semi-natural, moderate-to-fast-flowing, semi-upland river. Kingfishers and grey wagtails are quite frequent and most of Worcestershire's dippers are found. Common sandpipers have bred in the past and may do so in the future as there is suitable habitat and they breed upstream of Worcestershire. Lowland river birds such as sedge warblers and reed buntings are surprisingly scarce as there is apparently plenty of suitable rank herbaceous vegetation.

Mallards, which are usually breeding birds of still or slow-flowing water, breed only in areas where there is good bankside cover for nests. During the 1979 survey three post-breeding moulting flocks each of about 100 birds were found. These were in secluded areas with extensive bankside vegetation and exposed rock and shingle islands which provided safe loafing grounds. Mallards moult all their flight feathers simultaneously and require safe sites and ample food during the flightless moulting period.

Moorhens are generally not birds of fast-flowing rivers, but they were found fairly frequently during the 1979 survey, most often in reaches with more sluggish flow, silt deposition and clumps of emergent vegetation, usually bur-reed. Moorhens usually build nests within emergent plant cover over water, although at times nests are placed on tree and shrub branches overhanging water. Young moorhens, which resemble black cotton-wool balls, can swim soon after hatching and are led by the females to places where they can feed by collecting small insects from the water surface.

According to calculations made by J. J. Day following his survey, there were cumulative totals of 4,435 metres of steep earth banks, 5,020 metres of earth faces and 2,670 metres of marl or rock faces in the 57 km of river surveyed. There is, therefore, an abundance of suitable places for kingfishers and sand martins to excavate their nest tunnels. Kingfishers are fairly common in most years (about one pair every kilometre in 1979) although actual numbers depend on the weather conditions during the previous winter. Kingfishers feed mainly on small fish which they catch by plunge diving, so in icy conditions many die and the breeding population is low the following year.

Sand martins are one of the first summer migrants to arrive, with the first records at the end of March. They are gregarious birds throughout the year and nest in colonies often near water in river banks or sand and gravel workings. Horizontal nest tunnels up to one metre long are excavated into a suitable vertical bank of marl, sand or gravel. Sand martin populations in Britain are thought to have increased in the 1950s, when the many sand and gravel pits which were opened during the post-war building boom provided a superabundance of nesting sites. However, between 1968 and 1969 breeding seasons there was a sudden decline: by 1970 national populations declined by about 70% and in subsequent years fell still further to about 10% of the 1950s populations. It is now generally agreed the decline was due to drought conditions on their Sahel wintering grounds. Numbers have remained low ever since with perhaps slight recovery in recent years following small rainfalls in the Sahel. Sand martin holes along the Teme in 1942 and 1943 and 1976 were all between 330 and 360. Unfortunately, no counts were made in the 1950s when national populations were high, or in recent years. The river Teme is the best river for breeding sand martins in Worcestershire as there are only a few small colonies on the Severn and Avon. Quite large colonies can be found at several gravel pits.

Dippers are birds of rocky, fast-flowing streams with shallows and they are fairly common on parts of the Teme, mainly above Knightsford with highest numbers on the last 16 km of the Worcestershire river. Dippers are resident birds, remaining on their territory unless

Dippers nest along western rivers and streams. (RMB)

driven off by freezing conditions or floods. Along the Teme they appear to use tributary streams during floods, especially at the beginning of the breeding season. Dippers start to breed early in the year during February and March when suitable nesting sites under bank overhangs or on bridges may be inundated, although nests may be safe under an overhang and reached by the birds swimming underwater. Also, in flood conditions the underwater food supply may be unreachable. Dipper territories along the Teme are related to the presence of both shallows and tributaries and may encompass the lower reaches of a stream, where the nest is placed, and part of the river with shallows, which is used as a feeding ground in spring and summer. Dippers feed by entering the water in search of small freshwater animals, which are found when the bird flicks over stones with its bill. The main prey items are mayfly, fly and caddis fly larvae, with smaller numbers of stoneflies, although any small animals from freshwater shrimps to fish are taken. When feeding, dippers remain under water for about five seconds and may dive 10 times a minute. The insect larvae and nymphs which are the main prey reach maximum size in late winter and early spring at the time when young dippers are in the nest. The Teme and its tributaries are the main stronghold for dippers in the county. Simon Walker searched most of the county's fast-flowing streams and found between 40 and 50 regularly used nest sites, mostly in the west of county with a few in the north.

During the breeding season (mainly March to July) grey wagtails are primarily birds of fast-flowing upland streams and rivers, although they are found along lowland watercourses associated with stone-work (bridges, mills, weirs) and stoney shallows. They are found quite frequently along the Teme (in 1979 there were about 10-15 pairs), mostly above Knightsford, associated with shallows. Grey wagtails feed mainly on insects, both adults and larvae, which are common along watercourses — mayflies, caddis flies, stone flies, dragonflies, and ordinary dipteran flies are caught either by running and pecking or by aerial flycatching. The birds often run into shallows and catch anything that moves: tadpoles, small fish, insect larvae and crustaceans. During the 1979 survey many suitable unoccupied sites were noticed and population levels may have been low following the hard 1978-79 winter. Most grey wagtails leave their breeding grounds in winter and disperse over a wide area to places where insects are still available. In winter in Worcestershire they are often found foraging around farms with livestock, at sewage works, rubbish dumps and near lowland rivers.

The Laughern Brook joins the Teme at Powick about a mile from its confluence with the Severn, having followed a winding course from the north through the west side of Worcester. This brook is primarily a clay

Grey wagtails nest and forage on fast streams and rivers. (RMB)

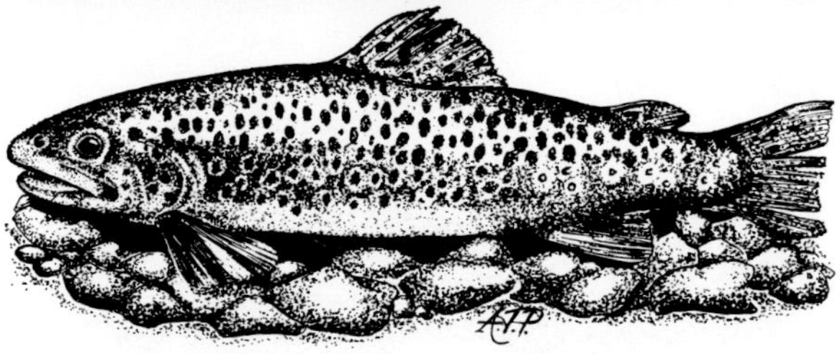

Brown trout inhabit fast-unpolluted streams and rivers in the west of Worcestershire. (WNCT)

stream, having many of the features of the Severn and the last part of the Teme. In particular the brook provides a useful wildlife corridor in both urban and country areas. Despite extensive dredging and clearing many natural history features have been retained, including pollarded willows.

Further upstream the next important Teme tributary is the Leigh Brook, which joins the river at Leigh, near a great tithe barn. This fast-flowing, shallow brook has many features of an upland stream. From Longley Green, where it is joined by the Cradley Brook, to the Knapp it flows through a narrow valley and is bordered by unimproved grasslands and woods. Part of the valley forms the Leigh Brook Valley SSSI, particularly at the WNCT's Knapp and Papermill Reserve. The first 12 acres of reserve near the Knapp House was established in 1968, the remainder in 1971. Sadly, opportunities to extend the reserve by woodland purchase in recent years could not be taken, due to financial restraints. The reserve is one of the most varied and attractive of those managed by WNCT, and is visited by over 5,000 people each year. The brook supports breeding dippers, grey wagtails, mallards, moorhens and kingfishers. During recent years dippers have disappeared as the brook has deteriorated through nutrient pollution by agricultural fertilizer, washed off cultivated land in the brook's catchment upstream of the reserve. Freshwater invertebrate populations have been severely reduced and both mayflies and the trout which eat them are now both much less common, which is unfortunate as the Leigh Brook was once one of the best small trout fisheries in the county. Trout can sometimes be seen climbing the weir, originally built centuries ago to provide water to power a mill, in use until 1973.

Butterbur — a stream-side plant of the west, but only male flowers are found in the southern part of Worcestershire. (LS)

The Papermill cottage on Leigh Brook — all that remains of an old mill. (LS)

At the other (west) end of the reserve there is a field containing the Papermill Cottage, which has given its name to the reserve. There are historical records

Siskin and redpoll feed on alder seeds in winter. (LS)

between 1677 and 1795 of a nearby fulling mill used to clean and dye cloth, and later used to make coarse paper. It fell into dereliction and was probably destroyed by a great flood in 1852, which damaged all the mills along the Brook.

There are many attractions on the reserve, including masses of marsh marigolds on wet ground by the pool near the weir in spring and beds of butterbur. The Brook is bordered by stands of alder, including a large pollarded tree near the weir. These trees grow on waterlogged soils and their roots, with nodules containing nitrogen-fixing bacteria, are often seen in the water at the edges of the brook. Alder sheds pollen from long catkins in spring and the fertilized female flowers form curious, small, woody, cone-like structures which contain small seeds. This is extracted for food by finches with small or finely pointed bills, particularly siskins, redpolls and to a lesser extent by goldfinches. Alder is the only British tree which relies on water to distribute its seeds.

The east and north side of the Teme valley differs geologically from the other side of the valley as the red marls of the Downton Series butt directly onto the Highley beds (grey clays, shales, sandstones, limestones and coals), part of the Coal Measures of the Carboniferous rocks, and there are no beds of Psammosteus limestone and therefore no tufa deposits. There are fewer tributaries on this side although those which do occur often form deeply cut dingles similar to the other side of the valley. One of the most notable is Dumbleton Dingle in which the Dumbleton Brook has itself cut a steep-sided valley down into the coal measures. Drift mines for coal were in use in this dingle at Hunthouse Wood until the 1940s and the spoil heaps are still visible: Hunt House colliery is marked on modern maps. The coal seams are part of the Forest of Wyre coalfield and coal has been extracted by shallow mining for centuries in the area around Pensax, Abberley, Mamble, Arley Wood and Shatterford. The presence of the old shafts and spoil heaps is often revealed by mounds covered with scrub and trees set in the middle of cultivated fields. The coal is generally of poor quality, sulphurous and in thin broken seams and no-one knows for certain how much has been mined. Despite these apparently poor prospects the district has been subject to a series of attempts to obtain planning consent for open-cast mining, so far all rejected. Dumbleton Dingle is an SSSI and most of it is covered by a WNCT 61 acre reserve at Hunthouse Wood. The stream in the depths of the valley is in deep shadow of mixed woodland and contains no plants. The damp undisturbed woodland is rich in mosses, liverworts and ferns. Open-cast mining would be extremely disruptive, one great concern being the release of toxic (heavy metals, arsenic) mine ground-water into the Brook, which joins the river Teme a mile away, and could severely damage freshwater ecosystems.

Dowles Brook in Wyre Forest — an unpolluted brook. (LS)

Soon after the Severn enters Worcestershire, it skirts Wyre Forest, which lies on the west bank, where it is joined by the Dowles Brook, which runs for about four miles through the centre of the Forest, from the confluence of the Mad Brook, Baveney Brook and Lem Brook at Furnace Mill. The whole of the Forest is of outstanding importance for natural history and is a National Nature Reserve and SSSI.

The native crayfish are in Dowles Brook now scarce, following onslaught by a fungus disease brought in with foreign crayfish. (LS)

The Dowles Brook has long been an excellent example of a clear, well-oxygenated, unpolluted trout stream. It curves and twists through a narrow, fairly steep-sided little valley, bordered by woods and small meadows of unimproved pasture. The brook is rich in freshwater invertebrates with abundant mayfly and caddis-fly larvae, and famous for its crayfish, which used to be extremely abundant, lurking under stones. Sadly, these native, lobster-like crustaceans, which can grow up to six inches long, have become scarce in the last two years. Virtually the whole population has been killed by a fungus disease originally brought to Britain from America on crayfish imported for breeding in lakes and ponds. The fungus lurks in mud and, if disturbance can be minimised for a year or so, the native may return. Crayfish are also found in small numbers in other fast-flowing streams in west Worcestershire.

The most abundant fish in Dowles are bullheads, also called miller's thumb, a large-headed spiney fish which spends much time on the bottom of the brook or under stones. Brown trout are common throughout the Brook and young chub abound. Eels, and occasionally lampreys, have been found.

Dippers, grey wagtails and kingfishers nest along the Brook, and mandarins can provide a colourful surprise for the streamside birdwatcher. They have a vertical take-off capacity and when disturbed lift easily from the water to disappear into the depths of the Forest.

The Stour joins the Severn at Stourport-on-Severn, which grew on the site of a hamlet called Lower Mitton after the Staffordshire-Worcestershire Canal was cut between 1766 and 1770. The river rises in the Black Country and drains parts of north Worcestershire and heavily industrialised areas of south Birmingham and Kidderminster: it is consequently rather polluted. It is medium-sized in its Worcestershire reaches and comprises about 10 miles of moderately deep, fairly fast-flowing river between the north county boundary and the Severn. Although the lower valley is floored with alluvium, higher up the river runs over soft Triassic Bunter sandstone. Parts of the river are 'canalised' with reinforced banks and there has been extensive dredging south of Kidderminster for flood control. There is little aquatic vegetation, restricted to fennel-leaved pondweed, the most pollution-resistant of species. Banks are often steep and emergent vegetation is scarce, with small patches of reed canary grass, bur-reed and reed mace. The Stour may be heavily polluted, but its natural history can be surprisingly varied. In summer Daubenton's bat and noctule bats hunt along many stretches between Caunsall and Stourport, the latter mingling with swifts late in the evening. Green sandpipers overwinter in quieter sections and the dock-strewn riverside meadows north of Kidderminster provide good chances of seeing a Worcestershire barn owl. Little owls yelp by day from old pollards and hobbies are quite frequent, especially in late summer, when young birds are attracted by swallows and martins massing prior to migration.

The Staffordshire and Worcestershire canal is about 46 miles long, with eight miles in Worcestershire. It carried goods from the Potteries and coal from Cannock Chase coalfield to Stourport power station. It runs alongside the Stour in Worcestershire and is similarly poorish for natural history due to pollution and disturbance. In the more northerly parts the canal carves a narrow course through sandstone cliffs which carry good growths of liverworts and ferns. Wall pellitory and black spleenwort are locally common. Ploughman's spikenard is found on steep canal-side slopes and tall emergent plants border much of the canal opposite the towpath, mainly bur-reed and common reed, with occasional clumps of reed mace, hemlock water dropwort, water figwort and valerian. Near Kidderminster the fiery orange flowers of jewel-weed, a balsam from North America, are common on rocky canal sides and old lock gates.

Along much of both the Stour and the canal there are extensive beds of Himalayan balsam, sometimes called policeman's helmet. This showy plant is a pest here, along parts of the Severn and its tributaries, and in many marshland areas. It quickly invades open wet ground and spreads, overshadowing many more interesting plants. It may grow to ten feet tall and form dense thickets. Seeds are 'fired' from ripe seedcases when they are touched and are carried by water and flood to new sites.

Both the river and the canal are of limited interest for wildlife but adjacent to them, near Kidderminster, are several important marshlands, wet meadows and areas of alder and willow woodland.

Two interesting brooks, the Dick and Shrawley Brooks, form the north and south boundaries of the parish of Shrawley, which lies on the west bank of the Severn about two miles upstream of Holt Fleet bridge. Both are trout streams and the longer Dick Brook lies in an attractive, winding valley with many small woods and occasional wild daffodil meadows. Andrew Yarranton, the engineer who worked on the navigations of both Stour and Avon, constructed locks on the lower part of the Dick Brook which enabled iron to be brought by barge from the Forest of Dean to a furnace and forge in dense woodland, presumably a source of the charcoal needed for smelting.

Among the varied flora of the Shrawley Brook are locally distributed species such as small teasel and giant bellflower. Less accessible banks are draped with the delicate fronds of soft shield fern.

The Salwarpe is a medium-sized lowland stream about 16 miles long, rising in the hills in the north of the county. The downstream half has some deep water with pools; the top half has many riffles and small beaches. Salt pits, pans and wells were present in Droitwich in Roman times. 159 salt pans are recorded in Domesday Book. The salt industry was important throughout medieval times, and a network of roads (the saltways), developed and radiated from Droitwich to carry salt over a wide area. Vast amounts of fuel were required to evaporate the brine and huge quantities of wood were obtained from the ancient Forest of Feckenham. Coal was first used in 1610 and Andrew Yarranton drew up a scheme to make the Salwarpe navigable to bring coal by barge to the salt pans. This was only partly successful and subsequently the Droitwich canal was built along the river valley, to be superseded by the railway in 1852.

Bullhead, a fish which lurks among stones in fast clean streams. (RMB)

At first salt was extracted from brine which appeared near the soil surface naturally; later it was taken from shallow wells and pits. The salt pans in Droitwich were all based in the Salwarpe valley and the river was naturally and unintentionally contaminated with brine. Vast underground salt reserves were found at Stoke Prior, about three miles north-east of Droitwich, in 1828 and this area eventually became the heart of a highly productive salt industry, alongside the Salwarpe and the Worcester and Birmingham canal. Deep extraction over 150 years or so led to surface subsidence and the formation of lakes and pools at Upton Warren, eventually important for saltmarsh plants and birds.

'Seashore' and 'saltmarsh' plants have been recorded in the area for centuries. After World War II the new Nature Conservancy included 60 acres of the Salwarpe valley near Westwood in its list of SSSIs, partly because of two obvious saline plants, dittander and wild celery, on the river banks, and partly because of the extensive reed bed which had developed in the then disused Droitwich canal. Since then the SSSI has been de-scheduled as part was used as a rubbish tip and this, together with landfill, the great expansion of Droitwich town, and more recently the re-opening of parts of the canal, has reduced natural history interest. Nevertheless both wild celery and dittander are still quite common near the small Briar Hill Coppice reserve, managed by WNCT for Wychavon District Council. Dittander is an uncommon plant of coastal saltmarshes and wet sand, and streams and ditches near the sea, mainly found in parts of East Anglia. Apart from a few sites where it was introduced, it is only found inland at Droitwich, where it was first noticed on the banks of the Salwarpe in 1853. It is a showy perennial plant growing to 150 cms with large clusters of small cream-coloured flowers. Wild celery is usually found on damp, brackish soils and in Britain has a mainly coastal (mostly in the south and east) distribution. The *Atlas of the British Flora* shows the species was distributed along the lower Severn in Gloucestershire and Worcestershire, presumably thriving before tidal flow upstream became restricted by locks. Other saline flowering plants found in the district include sea milkwort, reflexed saltmarsh-grass, common saltmarsh grass, sea spurry and the fleshy hastate orache, Babbingtons's orache, mud rush, cord grass, narrow-leaved pepperwort and club rush, plus *Enteromorpha intestinalis*, an alga of brackish water. Not all have been found in recent years and a new search and review would be of interest, especially as less brine is fed into the Salwarpe river or canal nowadays.

The Birmingham and Worcester Canal was opened in 1815 and is famous as the most heavily locked canal in Britain — 58 locks and five tunnels in the 30 miles from Worcester to central Birmingham. Feeder reservoirs at Barnt Green (Upper and Lower Bittel) and at Tardebigge are important for birds. The latter was excavated to provide clay to line the canal and it stands at the top of a great flight of about 20 locks leading down to Stoke works. Much of the canal is bordered by tall emergent vegetation, including extensive reed beds containing many reed warblers, sedge warblers and reed buntings.

There are many other streams in Worcestershire of considerable importance, such as the Bow Brook draining the centre of the county, the Hadley Brook near Doverdale with its accompanying string of small willow woods, and the River Arrow which drains from the northern hills near Alvechurch through Redditch and Warwickshire to the River Avon. They all have differing natural history features and are often a corridor for wildlife in intensively farmed land, urban and industrial areas.

LEFT: Severnside black poplar at Grimley — a rare tree of the Severn valley. (DMG) BELOW: The upper Severn — a naturally big river. (GHG) RIGHT: The great February 1990 flood inundated the flood plain of the Severn and Teme near Worcester. (DMG)

LEFT: Dittander, a saltmarsh speciality at Droitwich. (GHG) CENTRE: Reed bunting. (MH) RIGHT: Bogbean. (DH) BELOW LEFT: Palmate newt, (DH) and RIGHT: grass snake. (ARH)

ABOVE: The River Avon meanders through a wide flood plain. (DMG) LEFT: Bullrush is a typical emergent plant of the Avon. (DMG) RIGHT: The Avon is used by narrow boats and holiday makers. Water lily at the riverside. (DMG)

LEFT: The Birmingham-Worcester canal; frozen and quiet in winter but its reedbeds are nesting places for reed warblers in summer. (DMG) RIGHT: The marsh warbler — a scarce speciality of the Avon valley. (AWC) BELOW: Reed grass is a typical plant of moderately nutrient-rich brooks and rivers. (DMG)

LEFT: This old pollarded willow is intertwined with a sycamore which has grown from seed taking root in the willow crown. RIGHT: The shallows and gravel banks are typical features of the river Teme. BELOW: Clay cliffs along the Teme provide nest sites for kingfishers and sand martins. (All DMG)

Black poplars at Castlemorton Common. (DMG)

The Fringing Hills

The central Worcestershire plain lies in a saucer of low hills, open to the Severn valley to the south and to the north-east where the River Avon enters the county. These are not the dramatic scarps of northern Britain or the oak-clad hills of Wales. They have few unique species but they contribute greatly to Worcestershire's natural history.

The Malvern Hills are a prominent range some nine miles long, running south to north along the southern part of the old border between Herefordshire and Worcestershire. They are of great interest for their wide range of natural habitats, plant and animal communities, geology and use as a public amenity. A large part is a designated SSSI extending over 1,809 ha (732 acres), which includes unimproved acid grasslands on the tops, grading into taller herb communities lower down, some woods in the north and south, small areas of heathland, flushes, open water and bare rock faces in the quarries. Near Welland in the south another large SSSI, Castlemorton Common (158 ha, 392 acres) butts onto the hills.

The Hills and the adjacent land were declared a Royal forest after the Norman Conquest, carrying on a tradition already in place. It became known as Malvern Chase, subject to Forest Law designed to protect deer for privileged huntsmen, who prevented forest clearance, building and farming, although assarting was allowed for a fee. Eventually Charles I disafforested the land and sold a third. The remaining two-thirds were left open for the freeholders, tenants and commoners. In the late eighteenth and early nineteenth centuries many open areas were enclosed by parish Enclosure Acts and there were many

English stonecrop, found only on the Malvern Hills in Worcestershire. (WNCT)

private encroachments on the Hills. Eventually, under an Act of Parliament, 1884 the Malvern Hills Conservators were founded, to be responsible for commoners' rights and for maintaining the Hills for public enjoyment. This responsibility has continued to the present day in the face of ever-increasing difficulties posed by visitors. The tramping feet of a million a year causes path destruction, but the attraction is understandable as the views over many counties are well worth the climb.

Rising steeply from the east to 1,300 feet, the higher parts are clothed with the greatest expanse of acid grassland in the three counties of Hereford, Worcester and Gloucester. Dominant grasses are sheep's fescue, common-bent and wavy hair-grass with companion herbs sheep's sorrel, heath bedstraw and harebell. Crested hair-grass, upright brome, wild thyme, lady's bedstraw and mouse-eared hawkweed grow in base-rich areas. Rarer and more localised plants include upright chickweed, spring cinquefoil, bird's-foot, knotted clover and little mouse-ear. There are many disused quarries on the Hills, with disturbed soils, and these hold a number of rare and interesting plants such as fenugreek, buck's-horn plaintain and stork's-bill. In some quarries carline thistle, white horehound and common calamint can be found. Many of the lower slopes are dominated by bracken with woodland bluebells, wood anemones, dog violets and wild strawberry, flowering in spring before the bracken fronds unfurl.

Most of the woodland at the southern end of the Hills is sessile oak with birch and hazel in dry areas, and with ash and wych elm in flushed areas — a most unusual combination. There are wild service trees and both small- and large-leaved limes in some woods, with wood sorrel, yellow archangel and ramsons on the woodland floor, occasionally with locally rare climbing corydalis and violet helleborine.

There are a few patches of heathland with heather, bilberry, wavy hair-grass, cladonia lichens and mosses. A nationally rare clubmoss *Diphasiastrum x. issleri* occurs in a heathland fragment, the only site known in lowland Britain.

Wet flushes on Swinyards Hill contain jointed rush, lousewort, common spike-rush and marsh pennywort. Navelwort grows on rocky outcrops in the north, and its fleshy leaves are more common here than anywhere else in the county. English stonecrop is another local speciality.

The Hills are excellent for butterflies, with colonies of high brown, dark green and pearl bordered fritillaries, wood white and grayling. The caterpillars of all three fritillaries eat the leaves of violets and frequent rough, open areas associated with woods. Grayling butterflies are denizens of open grassland, with bare areas of rock or stone on which they sit with closed wings,

High brown fritillary, a butterfly of hilly woodland edges, now confined to a few sites. (LS)

almost invisible against the stone with their bodies tilted to reduce shadows. Their food plants are soft grasses including sheep's fescue. The species is now scarce at inland sites and something of a surprise to county newcomers.

The tree pipit sings in flight over rough grassland with shrubs. (WNCT)

A wide range of natural habitats supports a rich variety of breeding birds. Skylarks sing high over the hills and tree pipits in song-flight drop like paper darts above scrubby areas and near woodland edges. Meadow pipits, an uncommon breeding bird in Worcestershire, breed in the open grasslands, with small numbers of wheatears. Woodland breeding birds include wood warblers, some pied flycatchers and hawfinches. Breeding willow warblers are abundant and whitethroats common. One speciality is the regular autumn passage of ring ouzels which feed on rowan and other berries before moving south. Another is the regular late autumn occurrence of snow buntings on the hilltops. Ravens and peregrines are winter visitors and both have bred on the Hills in the past. Hunting kestrels hover or hang in the updrafts over hill edges, searching for mice and voles and family parties are a common sight in late summer.

Most small mammals occur on the Hills, which also support a large winter colony of lesser horseshoe bats. Dormice are found in some woods and there are records of polecats, a species which appears to be returning to the county with a scatter of sightings throughout the south. Superficially polecats resemble large ferrets and the two species do hybridize — the hybrids are difficult to distinguish.

Castlemorton Common consists of gently undulating land running up to the Hills. It lies on drift soils overlying impervious Keuper Marl and is covered with rough grassland, marsh and scrub. An intricate mosaic of plant communities matches the underlying wet and dry, acidic and basic soils. Patches of neutral grassland contain dog's-tail, red fescue, quaking grass, sweet vernal grass, and mat-grass but weakly acid areas are commoner with sorrel and tormentil. Stemless thistle and mouse-ear hawkweed grow in more base-rich patches. The drier areas of grassland are dotted with many active anthills, often with wild thyme growing on them, visited by green woodpeckers digging out ants to eat. The marshy grasslands contain tufted hair grass and soft, hard, conglomerate and jointed rushes grow along the streams, together with common streamside plants. Other locally uncommon species are marsh arrow-grass and broad blysmus, fleabane, common spotted orchid, sneezewort, marsh ragwort, marsh pennywort, lousewort, marsh bird's-foot trefoil and tubular water-dropwort. There are stands of parsley water dropwort, in one of its few Worcestershire sites.

Snow buntings are rare but regular autumn visitors to Malvern and Bredon Hills. (BW)

In one place there is a small pond containing *Ranunculus peltatus* and an uncommon little umbelliferous plant, lesser marshwort, which is insignificant in appearance but of great rarity.

Extensive stands of scrub contain western gorse, common gorse, hawthorn, blackberry and dog rose. Castlemorton's most unusual feature is its collection of about eighty pollarded black poplars. Hunched around ditches and small ponds, this gnarled gathering is one of the biggest concentrations in Britain.

Like the hills above, Castlemorton Common is an interesting place for birds. The rough terrain begs the presence of great grey shrikes and it is always a great ornithological thrill to see these handsome grey and black birds hunting among the scrub. Sadly, these superb winter visitors are occurring with less frequency. The Common's marshes attract breeding and wintering snipe in small numbers and reed buntings stutter their simple songs in places. The scrub supports whitethroats, yellowhammers, dunnocks and occasional stonechats and whinchats. Younger ears may also detect the monotonous reel of grasshopper warblers in the densest thickets.

To take in the view to the east from the Malverns on a clear day is to see Piers Plowman's vision of 'A fair field full of folk', with most of Worcestershire and parts of Gloucestershire and Warwickshire spread out beneath your feet. The 'Straits of Malvern' lie between the Hills and the Cotswold escarpment and the central plain of Worcestershire slopes gently up and away to the north-east and Stratford-on-Avon. Exactly to the east the dome of Bredon Hill rises to nearly 1,000 feet from the Avon Valley, with its steep north-facing slope crowned by a short tower. The south slope glides more gently into the Vale of Gloucester. Viewed from north or south the whaleback outline is less dramatic.

Bredon Hill is a Cotswold outlier and a geological sandwich of similar structure. The beds slope to the south and are capped by the Inferior Oolitic Limestone which forms the steep top slopes above Wollas Hall. Beneath are the mixed beds of the Upper and Middle Lias, and at the base the thick clays of the Lower Lias. A prominent feature is the top of the Middle Lias, where a bed of hardish, sandy limestone, the Marlstone Rock, is more resistant to weathering than the layers above and below, and gives rise to flattish spurs like Even Hill above Elmley Castle. Above this are the mudstones, shales and clays of the Upper Lias which, on the north side of the hill, are involved in a series of interesting earth- and mud-flows. The cap of oolitic limestone forms the summit north-facing scarp. This Cotswold Stone is, in the Cotswolds themselves, covered with thick beds of the Great Oolite. 'Oolite' means 'egg-stone' and when examined with a low-powered lens the stone is seen to be made up of small ovoids, rather like fish-roe, set in a hard structureless matrix. The land above St Catherine's Farm shows active and old mud-flows where debris from the slopes above collects in roughly bowl-shaped depressions and under suitable, often wet, conditions spills over the edge of the Marlstone Rock as a mud-flow across the lower land. Part of this area is now an SSSI and a National Nature Reserve for its interesting geomorphology and its limestone flora and fauna.

Bredon Hill lies next to the Vale of Evesham. This part of Worcestershire was cleared of forest and cultivated before the rest of the county. Bredon Hill was probably cleared fairly early on, since ancient woodland sites are relatively scarce. There were sheepwalks on the Hill in the 1600s and large areas of it were probably grazed as common land. In the nineteenth century sheep declined, cattle and horses were grazed on the upper slopes, and cultivated land crept up the lower slopes to the spring line, marked by the upper extents of the higher villages. During the twentieth century cultivation extended, mainly in the Overbury estate, to cover most of the southern slopes, with corn grown to the escarpment edge. Most of the higher northern slopes remained as rough pasture above the woods and these have a marked tendency to revert to mixed scrubland with hawthorn and bramble.

A good many plantations were also established in late nineteen and early twentieth centuries, including a beech wood which capped the hills for many years. This was cleared in the 1940s and the land cultivated. Larch plantations were planted on the eastern slopes. In the last twenty years attempts have been made to improve some of Bredon's north-facing rough pastures by levelling, scrub clearance and the use of herbicides and fertilizers. Whether this is a worthwhile exercise on such high and stony soils is questionable.

The natural history of Bredon Hill is similar to that of the Cotswolds, with many interesting plant and animals associated with limestone soils. The arable weeds include many uncommon in much of Worcestershire, such as Venus's looking-glass, corn salad, field madder, round- and sharp-leaved fluellens, dwarf spurge, green field-speedwell, fumitories, cut-leaved red dead-nettle and heartsease. Babington's poppies are quite common, identified by their yellow stem sap, and night-flowering catchfly is a speciality. The grasslands around the tower and on the nearby earthworks contain many interesting plants including pyramidal orchid, small scabious, field scabious, bee-orchid, quaking grass, wild thyme, salad burnet, fairy flax, rockrose, harebell, ploughman's spikenard, carline thistle and stemless thistle. One summer a solitary frog-orchid was found but none have been seen since. During days of sheep-grazing, there were great drifts of meadow saxifrage but this has decreased under cattle. Elsewhere other grassland plants are wild mignonette, sainfoin, horseshoe vetch (once the food plant of the chalkhill blue butterfly, but none remain), bird's-foot trefoil, purple milk-vetch, hairy violet and squinancywort (once used for curing quinsy). Among the rough pasture are many anthills, often with special plants. For example, spring whitlow-grass covers the meadow ants' hills with white flowers at the end of March above the 700 feet contour. Stoney outcrops are covered with parsley piert and crosswort is common on the north-facing stoney grasslands.

Bredon is a paradise for thistles. The agricultural pests creeping thistle and spear thistle abound in the rough pastures, while marsh thistles grace the wet flushes with seven-foot stems and purple flowers. Sit on the grass with care near the tower because the turf is replete with stemless thistles. In good years there are great droves of dramatically robust woolly headed thistle and thousands of purple nodding thistles, and the papery yellow and white flowers of the carline thistles can be found on the earthworks near the summit. The thistles themselves are biological playgrounds for many small insects and are well worth study for their gall-causers and their parasites, found both in flower heads and stems, and for leaf-miners. The patterned wings of picture-winged flies, whose larvae live in flower heads, are particularly attractive. Thistle flowers are visited by many insects, including butterflies and bumble bees.

There are old quarries on Bredon, both in the oolite and in brands of harder rocks in other beds, and these contain unusual plants such as twayblade, adder's tongue fern, long-stalked crane's-bill, hound's tongue, blue fleabane, autumn felwort (the nearest Worcestershire gets to a gentian) and yellow-wort.

The north-face scrublands are full of twisted, gnarled trunks of old hawthorns and ancient elders. The latter are probably the unpalatable survivors from the days when rabbits covered Bredon in thousands, to be virtually wiped-out in 1966 and 1967 by myxomatosis. Before the extermination, fields on the Hill were bordered by wide bands where the crop was destroyed by rabbit grazing. There are extensive bramble thickets and Bredon has been a happy hunting ground for generations of blackberry pickers. Wild rose tangles and mature hawthorns carry a heavy berry crop and the scrub is often alive with feeding redwings and fieldfares in the winter months. Many of the hawthorns are host to mistletoe, which is common in south Worcestershire. Whitebeams are found in some scrubby areas on the west of the hill.

Stoat family. (RMB)

Unlike the rest of Worcestershire there are old stone-wall field boundaries on Bredon, as on the Cotswolds, and these add an extra dimension. They are often lichen-covered and prominent species are the flat, white, paint-blob patches of *Lecanora calcaria,* the orange patches of *Caloplaca heppiana,* the chocolate-brown *Verrucaria nigrescens* and white *Verrucaria hochstetteri.* These lichens are grazed at night by the small hairy caterpillars of the muslin footman moth and during the day, even in the hot sun, by a small snail *Pyramidula repestris.* Calcareous Bredon and its walls are an excellent place for snails: John Meiklejohn has recorded at least twenty-four species and five species of slugs, mainly in quarries. Invertebrate excitements are also hidden beneath stones and in the walls. Eight species of woodlice have been found, including the tiny blind *Platyarthrus hoffmannseggi,* which is found in the nests of the yellow meadow ant. Four millipede species, centipedes and many beetles, including a big ground beetle *Cychrus caraboides,* which digs through the soil, hunting snails, have been found. Piles of empty snail shells reveal the activities of the snail-eating glow-worm larvae and the beetles can be found on summer's nights, glowing on the grassy banks.

The stone walls are also good for mice and voles, where they are hunted by weasels that flow through the crevices of the wall as if they lacked a skeleton.

Parts of the middle regions of the north side of the hill are covered with rather derelict open woodland, mainly of ash and hawthorn. Grazing sheep effectively prevent the growth of tree seedlings — many ashes are old and falling to pieces. They are often dotted with the black knobs of King Arthur's cakes, a fungus also known as cramp balls. Many of the hollow trees are used as nest sites by jackdaws, which fill them with piles of sticks to support a nest lined with sheep's wool. Tawny owls also use these holes for nests but may be turned out by jackdaws. Smaller holes are used by starlings and, until the national decline in the 1960s, redstarts were common — a few pairs still breed.

The steep slopes bearing the old ash trees are sites of large badger setts. Other common mammals are rabbits and a good population of hares survives, having disappeared over much of Worcestershire. A herd of fallow deer roams the hill, often causing damage to crops, but pleasure to visitors. Reptiles are generally scarce but slow-worms abound and grass snakes are common on the lower slopes.

Breeding birds include many skylarks, tree pipits and small numbers of meadow pipits. Like the Malverns, Bredon regularly attracts passage migrant wheatears, occasional wintering snow buntings and feeding kestrels. Buzzards are an irregular occurrence and, sadly, are still persecuted by foolish people in the name of game-preservation. In the 1950s woodlarks bred on the summit — the beautiful, lilting song of a woodlark high above Bredon is a memory to treasure. Alas, they are long gone from Bredon and most of England.

Broadway Hill is the small section of the long Cotswold escapment, which runs south into Gloucestershire and north into Warwickshire, that belongs to Worcestershire. The steep slope rises from Broadway village to the Beacon Tower, 1,026 feet above sea level. This imposing structure was built in 1797 by Lord Coventry for his wife so that he might view Worcestershire and his great estate at Croome. Reputedly thirteen counties can be seen from the top, including virtually the whole of Worcestershire, except perhaps the north-east and the land behind Bredon Hill. Round the tower is typical Cotswold limestone grassland and visitors can walk through grassy areas alive with marbled white and meadow brown butterflies. Many of the fields on the scarp slope are marked with ridge and furrow so were once cultivated, but have been sheepwalks for centuries and are clothed with short,

Marbled whites are butterflies of hillside limestone grassland. (WNCT)

herb-rich limestone grassland. The dominant grasses are sheep's fescue, red fescue, quaking grass, crested dog's tail, downy oat grass, erect brome and glaucous sedge. Herbs include bird's-foot trefoil, bulbous buttercup, hoary plaintain, mouse-ear hawkweed, salad burnet and burnet saxifrage with cowslips, thyme, common milkwort and harebells and occasionally meadow saxifrage on steeper slopes. Pyramidal orchid, green-winged orchid, rock-rose, kidney vetch and horseshoe vetch all occur more locally. There are some outstanding fields astride the public footpath from Broadway to the tower, with all these plants and many others, including small scabious and clustered bellflower. Notable are the summit quarries which have many twayblades, a few green-flowered helleborines and the scarce basil-thyme which colonises new spoil. Fellwort grows at Fish Hill picnic site alongside blue fleabane.

The Hill is good for butterflies and both small and chalkhill blues have been seen in recent years, due to the presence of their food plants, horseshoe vetch and kidney vetch. Common blue, meadow brown, wall, ringlet, small heath and large skippers are often seen. Grasshoppers are frequent in the unimproved pastures and, as on Bredon, there are many snails, woodlice and millipedes, often associated with stone walls. There are small beech woods and mixed broad-leaved woodlands on the Hill, some on ancient sites with extensive stands of dog's mercury and woodruff.

The West Worcestershire Hills run north from the Malvern Hills to Abberley Hills, crossed by the River Teme at the Knightsford Gap. The eastern end of the Abberley Hills is scarred by a large quarry. There are extensive ancient woodland sites on the hills, and pockets of interesting grassland, particularly on the Silurian limestones. WNCT's Penny Hill Bank reserve, near Martley, is a good example. This little bank bears most of the common limestone plants including autumn felwort, rock-rose, great-butterfly orchid and dyer's greenweed, a plant particularly associated with the Silurian limestones. More than twenty species of butterfly occur, including marbled white, green hairstreak, dingy skipper, brown argus. There are glow-worms here too. Other similar pockets of grassland have been found, sometimes encroached by scrub, occasionally with unusual plants such as large thyme, lady's mantle and adder's tongue fern. Walsgrove Hill, near Abberley, is a magnificent site for woolly thistle, bee-orchids and sweet briar at one of its few Worcestershire localities.

There are also interesting rocky outcrops. Osebury Rock, overlooking the Knightsford Gap, is surrounded by a typical western valley ash-wych-elm wood, mixed with sessile oak. This, like so many of the western woods, contains wild service tree and both small- and large-leaved limes and holly: yew grows on the cliffs. The Rock is Upper Carboniferous Haffield Breccia, with a mixture of volcanic rocks with limestone fragments set in a sandstone matrix. At the top of the Rock its acidic nature supports wood sage and greater wood-rush. The wet cliff faces have abundant common polypody fern and navelwort. The lower slopes are base-flushed and support ramsons, bluebells, nettles and yellow archangel. At the base are willow and alder trees. Many other wet, steep sites are rich in ferns and one bears the rare western polypody.

In the north of Worcestershire, rising behind the commuter village of Hagley, are the Clent Hills, forming a natural barrier between the Worcestershire countryside and the West Midlands conurbation. From the highest point at 1,035 feet, you can see most of Birmingham and the Black Country, while to the south and west the Malvern and Clee Hills dominate the skyline on clear days. The Hills are largley owned by the National Trust and 425 acres are managed by Hereford & Worcester County Council as the Clent Hills Country Park.

The proximity of the Clent Hills to a large human population makes them an invaluable recreation resource but the pressure of one million visitors each year inevitably takes its toll on wildlife. Natural history is further limited by geology. The well-drained acidic sandstone supports a restricted flora dominated by a few species but is ideal for growing certain conifers and large areas of Adam's Hill are clothed in plantations of larch and Scots pine. Bracken colours huge sections of both hills in autumn and is rapidly encroaching on the dwindling patches of open turf. On the lower slopes, small fields are given over to livestock and provide Clent with some of its richest habitats. Oak and hawthorn scrub occur with some regularity, interspersed with birch and rowan.

There are, of course, no completely natural landscapes left in Worcestershire, but Clent is more artificial than most. Perhaps the least affected are the remaining fragments of grassland, though eager botanists should contain their excitement. Sheep's fescue is everywhere, forming a dense sward which excludes all but the most tenacious plants. Wavy hair-grass, mat grass and sheep's-bit sorrel all manage to infiltrate the fescue lawns and in places the turf is starred with the tiny white flowers of heath bedstraw. Amphlett and Rea's *Botany of Worcestershire* published in 1909, suggests the grassland was once richer in species, citing records of pale toadflax (which still grows on walls in Clent village), and the greyish upright moenchia, a chickweed now apparently extinct. One sure loss is the stagshorn clubmoss, discovered in 1881 and uprooted by ' a rustic', who used the turf on which it was growing to cap a wall — literally a case of sod's law!

The enclosing fields on the northern slopes present a totally different picture. Flushed with neutral springs, they are excellent, albeit over-grazed, examples of old damp meadows. Adder's tongue fern grows in two small hay fields with common spotted orchids, quaking grass, heath grass and yellow rattle. Dyer's greenweed is widely scattered. Unpalatable to horses, its sulphur tufts vie with ragwort in otherwise close-cropped pasture. Wetter spots have fleabane, marsh arrow-grass and oval sedge, though tawny sedge, once frequent here, may have gone forever. Lady's mantle is a typical plant of these upland meadows and can even be found in St Kenelm's churchyard.

Clent's unplanted woodlands are restricted to the streamlines in the valleys and even these have undergone much re-planting. Brinks Coppice is one of the richest areas, harbouring a unique collection of county plants which include water avens, wood barley, and soft shield fern. Other woods are less rich, but larch plantations are better than pine and fir since, being deciduous, they let in more light. On Adam's Hill the larch wood understorey is typically elder or bramble with a ground layer of bluebells, wood sorrel and polytrichum moss. When Great Farley Wood near Romsley was felled recently, huge numbers of foxgloves sprang up and climbing corydalis appeared in hitherto unknown quantities. Fungi flourish under larches; at Nimming's Plantation the golden tufts of *Calocera viscosa* sprout from rotting stumps in autumn while in high summer the glades reek with the foul tang of stinkhorns. The larch woods are rich in birdlife, supporting breeding redpolls and willow warblers. Records of siskins in May are regular and it is likely they have nested, although verification is needed.

The most conspicuous broad-leaved trees on Clent are all aliens. Adam's Hill is crowned with a Mohican crest of beeches visible from Birmingham, each tree carved with a directory of visitors' names. Search carefully in the leaf litter in October and you may find a local fungus, *Boletus pruinosus,* which has a delicate bloom on its burgundy cap. Bramblings feed here on beech mast in good years and redstarts sing from the tops in spring.

Beech rarely reproduces at this altitude, but sycamore is extremely invasive. On Walton Hill groves of sycamore have become established to the detriment of native species. Their large leaves shade out the understorey and rot slowly, inhibiting the growth of woodland plants. Bluebells seem to be resistant, and the Hills boast some of the finest bluebell woods for miles around. Sycamores at Clent are particularly susceptible to infestation by tar-spot fungus which produces neat black blotches on the leaves. So close to Britain's second city, its presence is a vindication of the Clean Air Acts, for it cannot tolerate heavy pollution. No birds are specifically associated with sycamore, but mature groves like those behind the Vine Inn in Clent village are favoured by wood warblers.

Pedunculate and sessile oaks are well distributed, mostly as tall scrub scattered across the hillsides, with only a few large trees mingled with alien turkey oaks. Purple hairstreaks can be seen on hot days, flitting madly about the oaks, showing alternately light and dark as they flash their underwings.

The excellent drainage of the Hills means they have few boggy flushes. At one, on the southern flank of Adam's Hill, acid seepage prevents bracken from taking over. Here a collection of marsh plants includes star sedge, black sedge, marsh willowherb and square-stalked St John's-wort. Botanical absences are not always easy to explain: why, for example, is ling so scarce here when it is common at Kingsford and Lickey on either side? Reptiles are virtually unkown on the Hills: there are no lizards or adders and grass snakes are exceedingly rare. Is this because of public pressure or are land-use factors to blame?

Public pressure must affect bird numbers, but even so skylarks and meadow pipits are common, nesting among wavy hair-grass. Up to twenty pairs of the pipits breed, augmented by passage birds when flocks of over 200 have been recorded. Spring and autumn migration brings influxes of non-breeding species such as ring ouzels, which are an annual occurrence. Autumn birds can be detected among parties of thrushes in the rowans by their harsh, clacking calls. Wheatears have bred and may yet breed again on the quieter gradients of Calcot Hill; they are regular here on passage, investigating rabbit holes. Rise early and there is no telling what may turn up: one fortunate observer in September 1964 saw a dotterel on Walton Hill. Buzzards and hobbies are now annual and ravens have been seen on several occasions, bringing a whiff of the wild, Welsh hills into this distinctly urban upland setting.

East of Clent the high ground continues in a ridge across the Romsley and the Waseley and Lickey Hills. The steeper fields are grazed by sheep or horses and have lost much of their wildlife interest. Dyer's greenweed is locally common and downy rose crops up in the hedgerows in a few places. Great Farley Wood is largely planted with Scots pine and larch. Narrow and broad buckler ferns grow beneath, but the mountain fern noted by Amphlett & Rea has vanished, a victim not of coniferisation, but of the avaricious gardener. The same fate befell the oak fern, eradicated by collectors in the late nineteenth century.

Between Calcot Hill and Farley Wood lies Sling Common, a secluded valley threaded by fast-flowing Sling Brook. Dippers breed and the golden-ringed dragonfly has been seen, a scarce insect in the county. A number of flushes drain into the Brook and some have alternate-leaved golden saxifrage growing with its commoner opposite-leaved congener. The well-oxygenated stream contains miller's thumb and brown trout and provides a wintering site for water rails where it debouches into Sling Pool.

Waseley Hills Country Park is just south of Rubery and is managed by Hereford & Worcester Countryside Service, primarily as a green lung for Birmingham. It consists principally of rough pasture used for grazing cattle and has an uninspiring flora, typified by white clover, rye grass and crested dog's-tail. The large fields are separated by gappy hedges of hawthorn, holly and field maple, though the ranger service is doing much to improve the status of these hedgerows. The natural history interest of the Hills is centred on and around Segbourne Coppice, a sycamore and oak grove with a fine carpet of bluebells. Above the small pool is an acid valley lined with western gorse, grading into grassland dominated by wavy hair-grass and sweet vernal grass. Here there is more variety than on the Clent swards; harebells, black knapweed, musk mallow, and field scabious give a colourful show in midsummer. Among the gorse bushes, field pepperwort appears and there is even a record of greater broomrape for the late 1970s, though no plants have been found since.

The rich plant life attracts good butterfly numbers, all common species represented. Five-spot burnet moths are also abundant. Lepidoptera are in danger, however, from the scarlet caterpillar fungus, *Cordyceps militaris,* which comandeers their underground pupae, replacing the insect tissue with a tangle of fungal hyphae. Above ground, orange, club-shaped fruiting bodies testify to the struggle taking place below. Other Waseley fungi include the turquoise *Stropharia aerugnosa* and the multi-coloured parrot wax cap which can be yellow, green or red.

The Lickey Hills, lying hard against the Birmingham City boundary, are a valuable playground for local residents. Like the Clent Hills they receive enormous pressure from walkers, riders, picnickers and even orienteerers (if such a word exists!), though erosion here is less of a problem due to the harder Cambrian rocks which comprise the Hills. This is an altogether boskier habitat than Clent, with groves of birch and rowan alternating with thickets of gorse and broom. Mature Scots pine tower above clumps of bilberry on the steeper slopes, giving an almost Caledonian complexion to this somewhat crowded corner of the county. Yet there is none of the feeling of open space associated with Scotland and the wooded slopes of Lickey are dotted with large residences. Laurel and rhododendron mingle in the beech and fir plantations, creating an unmistakeable aura of suburbia.

Open spaces on the Hills are relatively limited and, inevitably, popular tobogganing venues in winter. In these areas bracken grows freely and was once, according to Amphlett & Rea, the centre of a localised industry: 'In years gone by there was a large local industry in the Clent and Lickey district in burning the bracken for the sake of the alkali in the ashes, out of which were formed round balls the size of a cricket ball, called "ess-balls", which found a ready sale in the neighbouring Black Country where they were used in the way of soap'. There must be many landowners in the county today who wish that some profit could be made from their acres of bracken.

One Lickey fern long since gone is the moonwort which, in 1907, was said to be flourishing near the Monument. Acid-loving hard fern still grows in wetter, more sheltered spots and is especially attractive in summer when it produces comb-like, sterile fronds. Nineteenth century Lickey must have shown more affinities to present-day Surrey than Worcestershire, as a scan through the 1909 *Botany* reveals. Edwin Lees, the eminent Victorian botanist, notes in 1856 that marsh violet and grass-of-Parnassus still persisted in boggy flushes, but were being badly affected by over-zealous drainage schemes. These were all too effective and both plants were soon eradicated. The same cause finally polished off the cranberries which Lees reports were made into tarts by a Birmingham confectioner. Modern Lickey has a limited flora characterised by ling and bell heather and any sites of botanical importance are guarded by a team of rangers, employed by Birmingham City Council to preserve the delicate balance between wildlife interest and conservation.

For such an artificial landscape, the Hills can boast an impressive roll-call of birds. A good morning's birding at the nearby Bittell Reservoirs is often topped off by a tour of Lickey by local birdwatchers and their efforts have provided some astonishing records. In 1975, for instance, Worcestershire's first breeding firecrests were found, when three of four pairs raised young. This promise was not followed up in subsequent years but the species occurs annually in the county and it is worth listening for the needle points of sound in spring, as long as you are confident of distinguishing this beauty from its commoner cousin, the goldcrest. Conifers have a bad reputation in the countryside but, when found with an admixture of broadleaves, offer a good diversity of food and shelter for many birds.

Tits can be particularly numerous at Lickey and populations are doubtless boosted by winter feeding by local residents. Coal tits, marsh tits, tree creepers and nuthatches are typical birds of the woodlands here, their presence especially obvious in winter, when they flock together for protection. There is also another hint of the uplands with breeding wood warblers, though singing birds have declined from twenty in 1969 to five at present, and even pied flycatchers on passage. Between 1949 and 1953 up to two pairs of this western quercophile bred. One lone male once sang vigorously from the chimney pot of Christopher Cadbury's previous house at Beaconwood, where there is an excellent oakwood. Nearby there are small fields with an abundance of pignut in the summer swards, with associated small, black, chimney sweeper moths. In irruption years, crossbills occur and have stayed on to breed in Scots pine and larch. There is also a breeding record of long-eared owl, whose bottle-blowing hoot whispers through spring plantations: investigating ornithologists should search carefully in places such as Lickey since proof of presence involves much prowling after dark — best to have a story ready for the local constabulary!

The polecat, thought to be extinct in Worcestershire, is now re-invading the county from the west. (BW)

ABOVE: From Bredon Hill, looking west to Malvern Hills. Note the slumping ground in the foreground — a feature of north Bredon slopes. (DMG) BELOW: Anthills on Castlemorton Common with the Malverns rising behind them. (DMG)

ABOVE LEFT: Kestrel and chicks. (ARH) CENTRE: Dyer's greenweed, near Romsley and RIGHT: woolly thistle. (Both DH) CENTRE: Abberley Hills, a western ridge of limestone. (GHG) BELOW: Grayling butterfly, camouflaged in a Malvern quarry. (DH)

ABOVE: The spine of the Malvern Hills and the central Worcestershire plain. (DMG) LEFT: Ancient hawthorn growing from oolitic rocks on the north Bredon escarpment. (DMG) CENTRE: Western polypody — a rare fern growing in the roadbank. (GHG) RIGHT: Sunken lane on Woodbury Hill with fern-covered banks. (GHG)

Stagheaded oak in dry farmland conditions. (DMG)

Farmland and Orchards

Farmed or cultivated land is the commonest land in Worcestershire, covering about 70% of the county. Farmland is what most citizens of Worcestershire see and drive through on their way to and from work. In 1991 few Worcestershire people actually work the land: most of those who live in villages are either retired or work in town. Few understand farming and many find it rather unfriendly and tend to associate it with nasty chemicals, stubble fires, hedge-flailing, tractors delaying main road traffic, vanishing footpaths, bulls in fields and big machines. Much is said about food-mountains, over-production, agricultural pollution, and many people worry about the effects of agricultural chemicals on themselves and wildlife. There is a general feeling that farmers care not a fig for wildlife. Farming has a poor image. This is unfortunate, although understandable, since farming is a successful and fruitful industry which helps supply those of us who work in towns with essential food.

To grow food efficiently farming has greatly changed in the last fifty years 'from an industry whose methods and systems were relatively favourable to wildlife to a specialised and highly technical business heavily biased against the maintenance of diversity of nature sought by conservation' (O'Connor & Shrubb, 1986). The major changes which have affected wildlife on farms in Worcestershire are hedge removal, improved land drainage, the increased use of chemicals (insecticides, fungicides, herbicides and, to some extent, fertilizers), the reduction in mixed farming systems and crop rotations, and the increase in large-scale, simpler regimes resulting in a loss of diversity. All these have reduced wildlife.

Mechanisation and the use of powerful machines have brought many changes. Previously intractable clays on the Lias and Keuper Marls have been brought into cultivation by the power and speed of modern tractors. Modern, fast machinery works best in a straight line at optimum speed and this has led to hedge removal on a large scale, to create more efficient, big fields. An estimated 2,550 miles of hedge have been grubbed up in Worcestershire since 1920, mostly in the last forty years. Additionally people using large, fast, powerful machinery are more distant from the land they cultivate. A man enclosed in the cab of a fast-moving tractor is unlikely to notice nests or a patch of a particularly interesting plant, and so avoid them, as he might have done in a more leisurely age. Another major effect has been to concentrate cereal planting in the autumn, rather than both autumn and spring. The resulting density and height of cereal growth by spring makes many crops unsuitable for ground-nesting birds, such as lapwings, which are now scarce breeding birds. In Worcestershire in 1989, 79% of the main cereal crops were autumn sown. Lapwings prefer spring cultivations for nesting so it is perhaps not surprising that less than fifty pairs were found in the whole county in a recent survey, whereas in the 1930s they were abundant. The soil disturbance caused by spring cultivation in the past exposed invertebrate food needed by birds at the start of breeding, and this is now much reduced.

Mechanisation in the last fifty years has led to the disappearance of working horses, which were numerous, and the consequent fall in demand for straw, hay and oats. This led to straw surpluses and straw and stubble burning, despite the development of short-stemmed cereals. Stubble burning has many effects besides the occasional obvious hedge and tree destruction which caused so many complaints a few years ago. The fires kill many small invertebrates (mites, millipedes, beetles, springtails) and have almost eliminated earthworms from fields burned annually, because of the loss of plant remains consumed by them. Straw-burning also reduces the amount of weed seed food for birds. It follows that hedgerow loss (removing nest sites), coupled with invertebrate and weed seed loss, have resulted in a steep decline in birds which feed on these items, especially finches such as linnets and goldfinches, and reed buntings, corn buntings and tree sparrows. Stubble-burning is of course in decline and will be banned, except in special cases, from autumn 1992. New techniques of chopping straw, disc-harrowing and rolling are being developed to ensure it will be rapidly incorporated into the soil. Unless the use of insecticides and other chemicals intervenes, this is likely to lead to improved soil structure and an increase in soil invertebrates.

Immediate post-harvest cultivation and autumn cereal sowing are fast operations with modern machinery. This affects birds by shortening the period when spilled grain and stubble weeds, their seeds and insects, are available. This and changes in grassland management are probably responsible for the decline in rooks. Rooks do best in areas of about half grassland and half arable as they obtain food from both: sown seed, spilled grain and grassland invertebrates are all important at different times of the year. Grass cut for hay is searched for insects but this resource disappears when it is cut early and taken as silage, and in any case numbers of invertebrates have declined in improved pastures and leys. Rapid cereal harvesting and sowing have further reduced an alternative food supply. Further loss of food arises from less spring cultivation disturbing invertebrates and reduction in weeds in winter. Put all these together and rooks move elsewhere! In central Worcestershire north of Pershore a 10 × 10 km square once contained about 1,000 rook nests, scattered through the square, mainly in big elms. Now there are a couple of hundred in two rookeries. The elms have also gone through elm disease but there are still plenty of trees, and the loss of elms is not responsible for the loss of rooks. Overall, rooks have declined greatly in the county and many old rookery sites are now deserted.

The use and sophistication of herbicides have both steadily increased in the last fifty years, leading to cleaner crops, although particular plants continue to give problems of control. This has led to loss of weeds, the insects dependent on them, and of the birds and small mammals which eat either the seeds, the plants, or the insects. The enormous decline in partridges stimulated much research. The main cause was an increase in the death of chicks due to a lack of insect food, itself dependent on weeds within cereal crops. The important weeds for partridges are knotgrass, black bindweed, mayweeds and charlock, all effectively

Lapwings are now scarce on farm fields but yellowhammers breed in many hedges. (RMB)

killed by herbicides. The dependent insects are sawfly larvae, plant bugs, leaf beetles and ants, all killed by insecticides and fungicides. The recent development and use of specially managed field margins and adjacent hedges to reintroduce weeds into parts of cereal crops, will regenerate other species and will perhaps save some such as corn cockle from extinction.

On arable land the hedge bottom may be treated with powerful, long-lasting herbicide to control weeds and thus be bare of plants. Or there may be a narrow band of ground between hedge and crop with a few weeds or tall herbaceous plants. Most crops are treated with a wide range of pesticides and a roadside hedge, if it is big enough, may prevent spray drifting onto wildlife-rich verges. The creation and maintenance of weedy fringes or headlands to arable fields is of interest to partridge shooters and conservationists alike. Such headlands can become refuges for arable weeds and insects, and cause little weed invasion of crops if properly managed, perhaps with a sprayed band between the weedy margin and the crop. There has been much research by the Game Conservancy and others on management of field margins and interesting practical work has been done by Adrian Darby on his farm at Kemerton, reported by Deane in 1989. Four main types of expanded field margins were used.

Conservation headlands are areas in which certain pesticides were excluded from the outer strip of cereal, encouraging annual weeds, including knotgrass, black bindweed, mayweeds and charlock. The sawflies, ants, plant bugs and beetles associated with these plants are important prey for partridge chicks, which survive well in these headlands.

Wildlife fallow margins are strips along the edges of arable fields which are cultivated but not cropped, so encouraging a diverse flora of non-invasive arable weeds. Poppies, hemp-nettles, mayweeds, fumitories are all weeds which attract specific plant-eating insects which in turn attract insect-eating spiders, beetles and shrews and their predators. Flowers like field pansy attract nectar-feeding butterflies. Ground-nesting skylarks may use the margin and the seed crop is eaten by finches.

Grass margins are strips of perennial grasses and flowers established beside both arable fields and grassland. Coarse swards, dominated by tufted hair grass, false oat grass and cocksfoot, give deep cover for small mammals and many invertebrates. Mown swards of perennial fine grasses and herbs create an edge similar to unimproved pasture or roadside verge.

Expanded hedgerows with scrub margins rambling into the field are aimed at enhancing hedgerow wildlife. All these methods increase wildlife at field margins and hopefully many more farmers will want to try them.

Both linnets and goldfinches have decreased in Worcestershire, especially in cereal growing areas. National studies have related declines to high nestling mortality, due to insufficient weed seed food. Linnets eat the seeds of several important crop weeds: chickweeds, polygonums and fat hen (especially in brassicas), all of which are vigorously controlled by herbicides. Although goldfinches have decreased, the effects have been less severe as they feed on thistle, dandelion, groundsel and similar plant seed, both in crops and on roadsides, woodland rides, rough pasture, and they take other food, for example alder, teazel and burdock seed. With its long, rather pointed bill, the goldfinch is the only British species which can extract seed from teazel heads. Linnets also take brassica seeds and, in the past, large numbers fed on the long seed pods of black mustard by the river Avon. Both linnets and greenfinches feed avidly on oil seed rape, perhaps some compensation for the loss of agricultural weed seed.

Between about 1964 and 1974 reed buntings, primarily birds of wetlands, moved into cereal crops as breeding birds and then declined. The crops chosen usually contained a good diversity of weed growth beneath the grain, providing a structure in which the birds could nest. After 1974 use of herbicides changed the favoured structure and reed buntings left the fields. In the 1970s many reed buntings were seen in corn fields, particularly between the Worcester-Pershore road and north to the Worcester-Alcester road. At that time the occurrence of corn buntings, yellowhammers, and reed buntings in smallish cereal fields and their hedgerows was striking. Corn buntings tend to be associated with barley rather than wheat. Yellowhammers, however, seem to do best in areas of mixed farming. They rear their chicks on insects, take grain in summer, and tend to move to fields or yards where stock is kept in winter, feeding on spilt grain and other food. The quail is another species which breeds in barley crops and in good 'quail years' the quick 'wet-mi-lips' call greets the early morning naturalist from the depths of the crop in May and June. A long series of quail records from an organic farm at Broome relates directly to the richness of the flora.

Arable weeds are an interesting collection of plants as they are adaptable and able to withstand harsh, disturbed conditions. Some of the less robust species are now scarce, as they have not been able to withstand mechanisation and the chemical regimes of modern arable farming. Buried seed may be long-lived and, given a chance, even the scarce weeds return in small numbers. For example, the little umbelliferous plant shepherd's needle, with its ridiculously long seed pods, was thought to be extinct until several colonies were found recently. Night-flowering catchfly, corn gromwell, corn buttercup (with spikey black seeds), the two fluellens, spreading hedge-parsley and knotted hedge-parsley have all been found by surprised botanists in recent years. Stirring long-undisturbed soil may have astonishing results. Sudden banks of scarlet poppies on roadworks are a dramatic reminder of the long-survival ability of weed seeds. A new roundabout at Droitwich was clothed one summer with corn spurrey, buckwheat and charlock and nearby an amazing carpet of several acres of mayweed appeared one year on newly tipped soil. All these were soon suppressed and boringly replaced by typical Council roadside grass, awaiting the next roadworks for a new blossoming.

Shepherd's needle, a scarce arable weed. (BW)

Common weeds on the heavy soils are couch, black grass, wild oats, annual meadow grass, knotgrass and sterile brome. Cleavers causes serious problems in crops, and chickweed, field speedwell, scentless mayweed and poppies are often seen.

The common weeds of the well-drained terrace soils of the market gardening areas near Evesham and Pershore include annual nettle, pennycress, the fumitories, groundsel, and scarlet pimpernel (opening its flowers in the sun and closing when clouds pass over). The seeds of fumitories and other low-growing weeds are important food for turtle doves, which nest in nearby tall hedges or scrub. In timcs past wool-waste or shoddy was used to fertilize terrace soils and many strange and alien plants from other countries grew for a few years. One of the most persistent is toothed medick, a native of Southern Europe.

The acidic fields of north-west Worcestershire overlie Triassic sandstones and quickly revert to heath when left fallow. At Larford, south of Stourport-on-Severn, a rich crop of arable weeds sprouts on the well-drained soils. Corn camomile and rough poppy may occur in such numbers as to colour the fields red and white. Annual knawel and bur chervil are abundant and an excellent find in 1990 was the re-discovery of smooth cat's-ear at its only known Worcestershire site. Tower mustard grows nearby. In 1986 a few plants of lesser snapdragon occurred, the first county records for many years. Other acid-loving arable weeds include corn spurrey and poppies. The rapidly increasing green nightshade, an alien from South America, is particularly common in sugar beet and potato crops. Corn marigold is found at Churchill and Iverley in most years.

Changes in weed populations are likely in the near future with the reduction in deep ploughing and the increase in direct drilling into relatively undisturbed soils, autumn sowing, and the use of vigorous, many-stemmed cereal plants (the successes of the plant breeders) and improvements in herbicides. Also cereal seed is becoming cleaner with few contaminating weed seeds. It will be interesting to see how the weeds respond. Some are increasing at present, including pansy, black grass and sterile brome.

In 1989 about 39% of Worcestershire was under arable: about 42% was permanent pasture and another 12% short-term (four years old or less) grass leys. Therefore 54% of the county was grassland. When presented with these facts a first reaction to the grassland figure might be 'where is it!' On reflection, one realizes that large parts of the river flood plains are grass, and there are extensive grasslands in the south-west, centre and north-east of the county. But with the decrease in dairy cattle, what is all the grass used for? The answer is probably more sheep, sileage and some cattle. From a natural history viewpoint virtually all these grasslands are 'improved' and relatively poor in wildlife diversity. There have been changes in the proportion of arable to permanent pasture in Worcestershire: in 1886 55% arable to 46% grass; in 1900 37% arable to 63% grass; in 1939 29% arable to 71% grass; in 1989 39% arable to 54% grass although, from an agricultural viewpoint, one might consider short-term grass leys as 'cultivated', in which case 51% of Worcestershire was arable and 42% permanent grass. However many of these 'permanent' pastures have probably been ploughed and re-seeded since 1945. In practice farmland in Worcestershire is about 50% each of arable and grass.

There have of course been great changes in the crops grown on arable land. In 1937, 30% of the land grew cereals: in 1989, 70%. In 1937, 16.5% of arable grew wheat: in 1989,

Old apple orchard at the Knapp, near Alfrick. Lichens grow on the old trees and fieldfares eat the fallen apples in winter. (LS)

38%. In 1937, 1.2% grew barley: in 1989, 26%. Also in 1937 about 46% of the cultivated land was used for fruit, market gardening and hops: by 1989 this had fallen dramatically to about 5%.

The disappearance of orchards has obviously changed the face of the county, famous for fruit growing for hundreds of years. In the 1940s people visited Worcestershire in spring to follow 'blossom routes' to admire the acres of plum trees covered with white bloom. A glance at the maps from the turn of the century shows orchards everywhere; plums were an important export to big towns and huge quantities of Pershore yellow egg plums were harvested until the 1950s. Apples and pears were also widely grown, for cooking, dessert and cider and perry making. Worcestershire was the centre of a flourishing perry making industry in the 1700s. The small perry pears grow on big trees, and many of those in the county were fully grown in the 1700s. The remnants of these old orchards can be seen scattered about: in 1990 they are decaying and dying of old age, but many can still put on a magnificent display of blossom in spring. Cherry orchards are a speciality of the Teme valley and visitors still go to view the blossom. The ripe cherries attract marauding blackbirds and starlings and even secretive hawfinches with massive bills. They alone among British birds can crack cherry stones to extract the kernel.

Remaining old orchards are now declining into senility, often an interesting state for naturalists as the old trees support a wealth of wildlife. Apples and pears are usually the best: uncommon lichens grow on old trunks; the old hollow branches are used as nesting tunnels by little owls and stock doves; the branches may be weighed down by clumps of mistletoe, and the dead wood is home to rare beetles. Where old orchards still survive, some have interesting grasslands beneath the trees, carpeted with cowslips and rough with anthills. Fallen apples attract large winter-feeding flocks of fieldfares, redwings, blackbirds and other birds. Across the county great acreages of plum, apple, cherry and pear orchards have vanished and many an ancient variety of cider apple is now only found (hopefully) in a special collection of old breeds. Some new apple orchards have been established, especially near Holt, and recently, a few acres of plums.

Hopyards are a feature of the Worcestershire-Herefordshire border, mostly near Bromyard. In 1800 there were 6,000 acres of hopyard in Worcestershire, explaining the demand for long, straight ash and lime poles. The lakes in Shrawley Wood were once hopyards. Demand declined and by 1937 there were about 2,000 acres. Currently there are about 1,000 acres near Bromyard and new yards can be seen near Worcester. There is great fear of the killer disease verticillium wilt, for which there is no cure and, as there is great danger of people carrying the disease into hopyards, please keep out unless you have express permission.

The market-gardening region of the Avon valley, near Pershore and Evesham, has changed like everything else. In 1937 the old characteristics were in place. Market gardeners were divided into roughly two types: smallholders with 5-15 acres sited mainly east of Evesham, and large-holders with 20-50 or more acres mainly near Pershore. Large-holders tended to grow big acreages of perhaps, sprouts, cabbages, bush fruit, or beans whereas a smallholder would grow anything up to fifteen different vegetable crops and a few soft fruit bushes and plums. A smallholder's land was often scattered in strips over a wide area. Many people worked on the land, for example 875 acres at Offenham were worked by 337 growers, usually helped by their families. This led to an almost medieval landscape, as around Badsey, with innumerable cultivated plots and no hedges. In 1937, land usage east of Evesham was divided into three roughly equal main parts: grass, fruit and market gardening. West of Evesham about 45% was grass, 18% fruit and 17% market gardening. In the east, famous crops were sprouts and asparagus; in the west, sprouts and runner beans. 73% of the orchards were plums, and about 17% apples. The grasslands were, and still are, the Avon flood plain meadows.

In more modern times the size of holdings in the Vale of Evesham has increased, land has been turned to cereals and other agriculture, and there has been an increase in salad crops under glass and polythene. The view from Windmill Hill, the WNCT reserve above Offenham, is still one of intensive market gardening, shimmering with polythene in early spring. Here and elsewhere the plums have gone. Perhaps the greatest changes have been west of Evesham and these can be appreciated by the modern summer view to the north from Bredon Hill, when most of the landscape is yellow-brown with ripening cereal, broken only by the green band of grassland along the river Avon and the roads and villages. At Bricklehampton is a vast cornfield where in 1937 there were twelve mainly market gardening holdings.

Generally speaking, farmland in Worcestershire is fairly poor in wildlife, although much depends on the intensity of cultivation. A few surprising pockets of old-style pre-war farming still exist. Changes can be expected in response to new demands and manipulative economics. Some land may be set-aside or managed for less intensive farming. Some farmers will take advantage of the call for 'organically grown' produce and use systems less aggressive towards nature. Land at the Worcestershire College for Agriculture is being used to teach organic farming and research in Denmark has shown that organic farmland is richer in wildlife than other farmland. New crops are being introduced and of course the brilliant yellow of oil seed rape has hit everyone between the eyes in recent years. About 3,000 ha were grown in Worcestershire in 1989 and this is likely to increase. It is being joined by large blue fields of linseed, a plant grown long ago for flax, and attacked by flax dodder.

FWAG, the Farming and Wildlife Advisory Group, has been fairly active in Worcestershire, with some farmers committed to making space on their farms for wildlife. Their work has included improved hedge management, planting new hedges and small woods, making ponds and wetlands, and conserving old pastures. WNCT wishes more farmers would become involved in these activities. Farmers and landowners interested in hunting and shooting maintain suitable areas for partridges, pheasants and other quarry and these support other wildlife, although accompanying gamekeeping can be detrimental to birds of prey, despite protection by law.

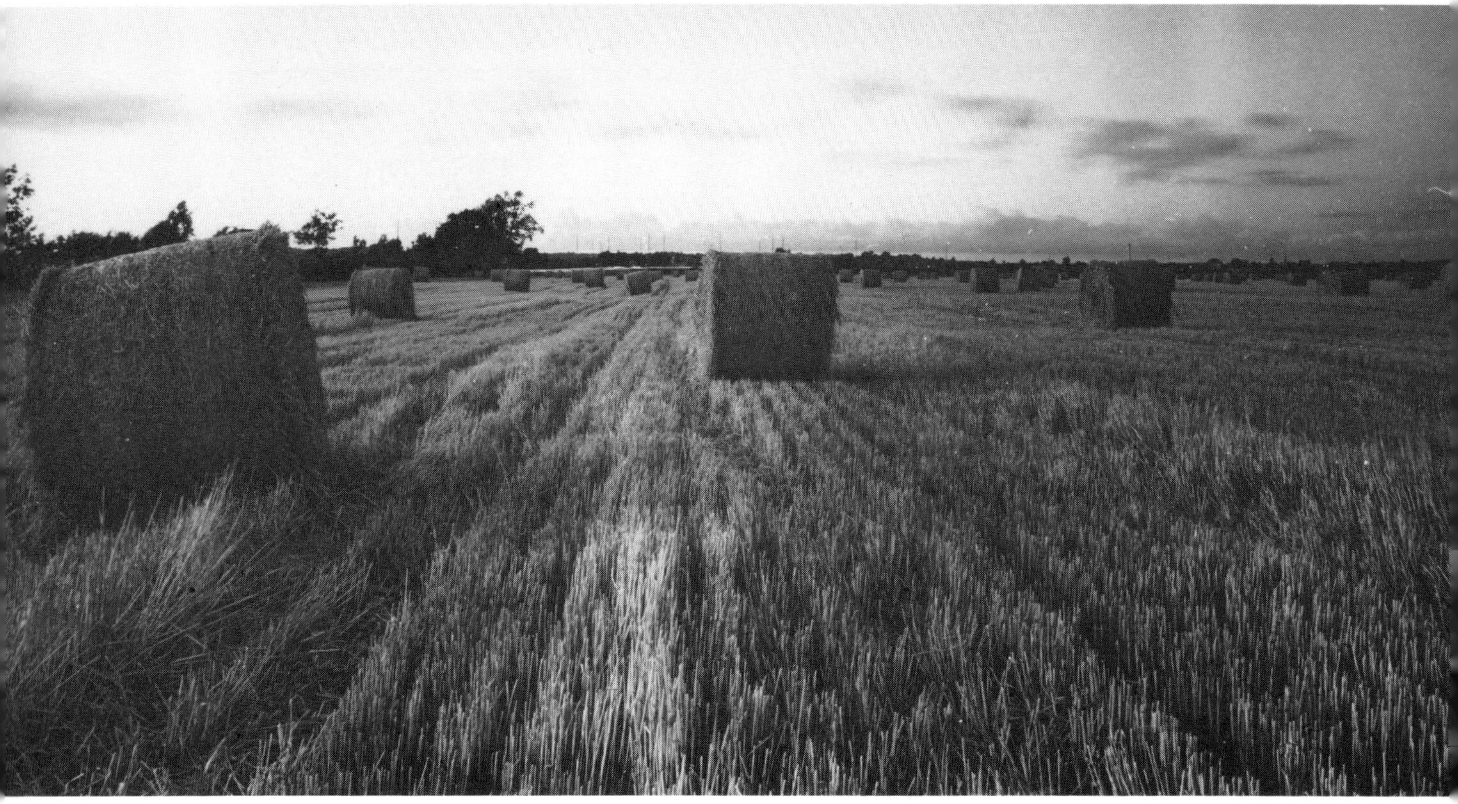

ABOVE: Medieval ridge and furrow are clearly revealed by snowfall. (DMG) BELOW: Modern harvesting — wildlife is scarce in these landscapes. (DMG)

ABOVE: Lone tree at harvest time. (DMG) BELOW: Modern irrigation in intensively cultivated land. (DMG)

ABOVE: Rooks are not so common nowadays in most of Worcestershire. (DMG) BELOW: Runner beans in polythene frost shelters — intensive market gardening. (DMG)

Old orchards are becoming scarce. Often used for grazing, they may contain a wealth of flowers. Little owls nest in holes in the old trees. (DMG)

Cow parsley and ash tree with heavily cut hedge in a landscape of oil-seed rape. (DMG)

Hedges and Verges

The view from any vantage point in Worcestershire shows miles of hedgerows dividing the countryside into fields. Fields with boundaries of hedges, walls, banks or ditches have been present in parts of Britain since Iron Age times. The Romans created field systems and in places these or even older boundaries are cut across by Roman roads, revealing their antiquity. The tradition of field enclosure persisted into Anglo-Saxon times and, in what Oliver Rackham calls 'planned countryside', were often obliterated by medieval open field systems. Eventually the great open fields were once more divided by hedges, often at the time of the Enclosure Acts in the late 18th and early 19th centuries. In Rackham's 'ancient countryside' the tradition of small enclosed fields persisted and in some areas they have survived. During the last fifty years many miles of hedgerow in both ancient and planned countryside have been removed, disrupting field patterns which have, in some instances, existed for 1,000 years.

In Worcestershire most field boundaries are marked by hedges rather than stone walls, although the latter are seen on Bredon Hill and on the Cotswold slopes near Broadway. Hedges and walls, although marking boundaries, are primarily intended to fence fields to contain farm animals. In modern times post and wire fences are often used to avoid the expense of repairing walls and maintaining hedges. For hedges to become and remain effective barriers to stock they have to be 'laid' or 'plashed'. When new hedges were planted during the enclosures they were protected for a few years from grazing animals by temporary fences, perhaps using willow rails, which were abundant and cheap (but soon rotted), until they had grown big enough to be laid. A hedge is laid by cutting through about three-quarters of the thickness of the stem near the ground and pulling the bush over nearly parallel to the ground, in the direction of the hedgeline. By pulling down one bush onto another laying proceeds until an impenetrable barrier is formed. The bushes are kept in place by a line of short hedge stakes and the whole is bound together along the top with long heatherings, usually of hazel woven round the stakes. The portion of remaining living stem keeps the bush alive and both the horizontal stem and the base of the plant re-sprout vertically, so reinforcing the barrier and providing the new branches for future re-laying. Side and top shoots can then be trimmed as required for many years before the hedge requires re-laying.

This method of hedge management, laying and constant re-invigoration, will keep the hedge alive and effective for centuries. Apart from some outstanding exceptions, hedge-laying in Worcestershire is out-of-fashion mainly because it is skilled, labour-intensive and therefore too expensive. On WNCT reserves the work is often done by volunteers who gain great pleasure from learning the skills of such an ancient and useful craft, despite the pain inflicted by spiny hawthorn.

Hedges may become established in three ways: by planting; as ghosts of original surrounding woodland, and by unintended colonisation of fence lines by scrub. Hedges planted to demarcate Enclosure Act fields were usually of hawthorn. Pitt, writing in 1807 of the results of agricultural surveys in 1794 and 1805, states that 'modern enclosures are made of post and rail, and sometimes with two rows with mounds, and a quickset fence planted between. The modern quicksets are the white hawthorn without any admixture of smooth wood; but some mix crabtree and holly'. Plant nurseries grew vast quantities of 'hawthorn quicks' and, for example, according to Gaut, in 1805 Messrs Harvey of Elmley Lovett had seven million transplanted five-year old plants for sale. Again, Pitt tells us that vast numbers of quicks could be produced by root cuttings. Quicks raised from berries were ready in three to four years and, before planting, their roots were pruned and the bunches of roots planted in 'rich mould' where they soon sprouted. Pitt also states that 'The greater part of this county is ancient enclosure, the fences often being full of timber trees, particularly elm, of which this county produces the finest in England; the fence [hedge] itself is here often composed of smooth wood, as elm, willow and hazel, as well as of crabtree and hawthorn'. In many parts of the county, fruit trees were often included.

The hedges of ancient countryside may or may not have been planted. Rackham quotes historical documents referring to a hedge planted in 940 and, at Grimley in Worcestershire about 966, of a hedge being destroyed. He points out that there are several Old English words for 'hedgerows', 'hazel rows', 'thorn rows' and 'willow rows' and references to 'old' hedges written in 816. These references suggest that hedges were probably planted over 1,000 years ago. Hedges were also formed by utilising trees and bushes already in place when fields were cut out of ancient woodland, thus incorporating woodland species like small-leaved lime, spindle and wild service tree into hedges. This was carried to extremes when a complete wood was grubbed-up, leaving a band of original shrubs and trees round the edge of a new field. The irregular ghostly outlines of old woods can be seen on maps or aerial photographs. A good and more recent example is visible at WNCT's Hornhill Wood reserve, near Himbleton, where the hedge surrounding a small field to the north-east of the wood was once clearly part of the wood.

Hedges unintended by invasion of a fence-line by woody species, often in times of agricultural depression or neglect, can occasionally still be seen in embryo in the Worcestershire countryside. Hedges often develop in this way from invading scrub alongside railway lines. In damp areas crack willow stakes sprout readily, forming large trees in only a few years.

The study of the history of hedges was stimulated by Pollard, Hooper and Moore and the book they wrote in 1974. Their research related the actual age of a hedge, determined by historical records, to the number of woody species found in a 30 yards length. Hooper's underlying contention is that more and more woody species invade hedges as time passes, roughly at the rate of one new species every 110 years. It is also probable that more species were actually planted in ancient hedges and that most enclosure hedges were of hawthorn alone. Further research has shown there to be many exceptions to Hooper's Rule, which cannot be generally applied. Max Hooper's great achievement has been to stimulate the study of hedges, to raise people's appreciation of their long and interesting history and to encourage hedgerow conservation. Before 1974 few cared about their fate.

Hedgerow field maple with winged seeds. (LS)

Few detailed hedgerow studies have been published in Worcestershire. In 1981 Wendy Johnson wrote a short booklet on the subject for WNCT, partly based on a study of hedges near Inkberrow, in which she showed that hedges of known different ages were perhaps not accurately aged by the Hooper method. For example, a possible Saxon hedge at least 1,000 years old had a calculated age of 470 years; and 'young' hedges about 100-160 years old had calculated ages of over 300 years. Interestingly some of the hedges in an area marked 'old enclosures' on the 1817 enclosure map had a calculated age of over 500 years and contained many (nine and ten) woody species, including woodland species — spindle, dogwood, woodland hawthorn and hazel. These hedges may have been formed when the original enclosures were cut out from woodland or have been invaded by nearby woodland species. The Saxon hedgeline overall contained similar woody species (ten and eleven) and also plum. All the calculated figures are estimates and, statistically, in 95% of cases the actual age is predicted to be within plus or minus 200 years of the value given.

The Enclosure Acts, which divided commons, common fields and wasteland into new field systems, affected parishes throughout Worcestershire, mainly between 1761 and 1820, but continuing to 1865. There were two sorts of enclosure: one to commence first-time cultivaton of boggy land, wastes, and heathland; the other to convert large strip-cultivated common fields to arable and grassland fields. As mentioned by Pitt, much of Worcestershire was already ancient enclosed land — the 'ancient countryside' — and Gaut gives three examples to show the varying proportion of old to new enclosure at the time of the Act. At Hadzor there were 700 acres of old enclosure and 133 acres of unenclosed common; at Great Hampton, 345 acres of old and 806 acres of new enclosure; at Little Hampton, 33 acres of old and 404 acres of new enclosure. Records from the early fourteenth century, quoted by Buchanan in 1944, show that assarts (forest clearings) were numerous at that time, with many instances on heavy claylands near Alvechurch. Between 1760 and 1870, common fields and common land totalling 18.1% of Worcestershire had been enclosed under the Acts. In 1874, 4,253 acres of common land in 20 parishes remained. Between 1880 and 1940 the pattern of countryside set at the time of the enclosures changed little.

In Worcestershire the difference between ancient and planned countryside can be seen and even now affects the natural history of a parish, despite recent severe change. Rackham says that Anglo-Saxon charters show that hedgerows were customary in most of Worcestershire except in the south-east, which seems to have been the part of the county first cleared of woodland for cultivation in Roman or pre-Roman times. Around 1600 the south-east was mainly planned countryside where 85-95% of the land had open fields and commons. In the central band of the county running south west to north east the same aplied to 30-50% of the land, in the north west to 50-70% and in the extreme west (Teme valley and Wyre forest) to under 30%. Nowadays the south-east is the most open part of the county and the north-west the most wooded. The transition from planned to ancient countryside between the south-east and central Worcestershire can be appreciated both from maps and on the ground. North of Naunton Beauchamp the fields are smaller, the hedges less straight, there are more woods, the small roads are rarely straight and often sunken between banks, and green lanes are frequent.

A random examination of fifty-nine 30-yard lengths of roadside hedge made in 1989 gives some idea of the woody and climbing plants found in Worcestershire's hedges. The surveyed hedges were in three districts: the south-east on basic lias clay; on the central Keuper marl, and on Keuper sandstones with well-drained, often acid soils near Hartlebury.

ABOVE LEFT: Yellowhammer and (ARH) RIGHT: wood mouse. (DH) LEFT: Hornhill Wood, showing coppice and 'woodland ghost' hedgerows CENTRE: Meadow sweet and rosebay willowherb, Clent and BELOW: dropwort on the verge at Crossway Green. (Both DH)

The number of woody species per sample length ranged from one to eight with an average of four. They were (% of hedges in which species was found) hawthorn (85%), blackthorn (40.7%), field maple (37.3%), English elm *Ulmus procera* (37.3/), hazel 27.1%, ash (25.4%), elder (22%), oak (18.7%), privet (15.3%), holly (13.6%), plum (10.2%; dogwood (6.8%), crab apple, sycamore, wych elm, cultivated apple (all 3.4%) and small-leaved lime, yew, spindle and sweet chestnut (all 1.7%). The climbers were blackberry (71.2%), dog rose (49.2%), ivy (22%), black bryony (18.7%), honeysuckle (10.2%);, woody nightshade (8.5%) and white bryony (8.5%). One 'hedge' consisted of Japanese knotweed, an aggressive garden escape. Most hedges contained large amounts of hawthorn, blackberry and dog rose. Several samples contained only one species: elm, hawthorn or plum. The plum had obviously been planted and the pure hawthorn hedges had been planted recently during road widening schemes. Many hedges, especially in the south-east, contain long stretches of English elm which suckers strongly and suppresses other plants in hedges.

Any indigenous tree and scrub species may be found in a hedge and in west Worcestershire it is common to find wych elm, wild service tree and small-leaved lime, all woodland species. Lime is most frequent in the 'lime zone' west of the Severn and centred on Shrawley. Large-leaved lime, an uncommon native, grows in hedges near Suckley. Holly is also common in the west and north on the well-drained sandstones, where it was probably planted for its Christmas berry crop and for timber as well as a hedge. Holly hedges are often closely trimmed, but fine old trees line some westerly lanes. Yews are also found in these areas. Hazel, a coppice plant of underwood, is found in hedges in the ancient countryside of central and west Worcestershire, perhaps as a remnant of woodland or a woodland edge. The hedges bordering the road at Goosehill Wood, near Himbleton, are mainly of wild service and hazel. Buckthorn is locally abundant, particularly in some of the Avon flood-plain hedges, and is the food plant for brimstone butterflies. In sunken lanes the hedge banks are often massive tangles of old stumps and ivy.

Old man's beard or wild clematis, a vigorous climber, is common in limestone areas where it may festoon hedges and cover them with the large, downy seed-heads in autumn. This plant indicates places where the soil is limey and these areas are well worth examining for lime-associated plants. Other hedgerow climbers include wild hops, black bryony and white bryony, the latter with its spring-coiled tendrils, the former climbing by twisting round stems like runner beans. Bryony has large root tubers which are poisonous to cattle and apparently addictive — if a cow eats bryony root and recovers it will go and look for them when returned to the field.

Hedgerow trees were once an important source of timber and, in the ancient countryside of central Worcestershire, the many hedge trees still give the landscape an enclosed and more intimate feeling than in the relatively treeless countryside of the south-east. Pitt particularly mentions elm as the 'Worcestershire weed', a term also used in Warwickshire. In fact both counties used to grow majestic elms in enormous numbers in the days before the current epidemic. Elm was widely used for making the main framework and planking of farm buildings, for water pipes (elm rots only slowly when immersed in water), as an alternative to oak in timber-framed buildings, and for making coffins. Elm disease killed millions of trees in Worcestershire: an occasional big wych elm or Huntingdon elm can still be found but most elm has survived as scrub in hedges or small trees in woods where suckering root systems continue to produce a mass of new stems. Elm was most abundant in south-central Worcestershire hedges and the remains of massive trunks 3-4 feet in diameter can still be seen, though many have now rotted away. Small elm copses were common adjacent to hedges where suckering elm had invaded neglected land. In the 1970s, when hedgerows were full of dead and dying elms, there was a noticeable increase of great-spotted woodpeckers breeding in farmland areas, benefiting from a superabundance of wood-boring insect larvae in the dead wood.

Elm disease is caused by an ascomycete fungus, *Ceratocystis,* which blocks a tree's system for conducting water and nutrients. The fungus enters the tree, and travels from tree to tree, aided by elm bark beetles *Scolytus scolytus* and *Scolytus multistriatus*. Beetles lay eggs in a gallery excavated between bark and wood of recently dead elm. Hatched larvae excavate galleries under the bark and eventually pupate for the winter. The fungus, already present in the dead tree, forms fruiting bodies and spores in the galleries, which contaminate the beetles when they emerge between May and October. The beetles fly to living elms and feed by nibbling bark and wood in the angles of twigs, causing minimal damage but introducing fungus spores and thus the fungus into the tree. The growing fungus obstructs the conducting system and branches and twigs may die immediately. The tree may survive, or the fungus may invade the whole tree, killing it entirely. If the fungus enters the conducting system within the roots of a group of interconnected trees, which have grown from suckers, it may quickly kill a whole group of big trees. The current epidemic, which has killed most big elms in southern Britain, started in 1965 and is caused by a new and 'aggressive' strain of fungus, which has killed trees more quickly than in earlier outbreaks. The last epidemic in England occurred in the early 1930s and killed 10-20% of mature elms. It peaked about 1936 and then declined. Many elms grew again by suckering and formed young trees in hedges during a time of depressed agriculture. Historically, there appear to have been a series of outbreaks of elm disease, but none as serious as the present one. In prehistory, elms became abundant in post-glacial times and the pollen record shows they were one of the most common trees in the Atlantic Period between about 5,000 and 7,000 years ago. At the end of that period the elm pollen was suddenly reduced by half — the great 'elm decline'. Rackham postulates that this was due to a combination of elm disease and an increase in agriculture, which opened up woodland and allowed a wider and more rapid dispersal of elm bark beetles to infect many trees in a short period.

Will Worcestershire grow elm trees again? Nowadays young suckering elm is usually re-infected when it reaches a height of about fifteen feet, and it dies back. In hedges modern flail cutting rarely allows trees to reach a fraction of this height. In woods much the same pattern seems to occur and in WNCT woods attempts are made to coppice woodland elm so keeping it alive until disease-free days return. Until the disease itself loses virulence the return of big elm trees seems highly unlikely.

Other common hedgerow trees in Worcestershire are oak, ash and willow, probably in that order of abundance, although no counts are available. Willows are usually grown as pollards in wet areas and in times past used to provide a regular crop of poles. Ash and willow provide wood for split hurdles and ash was used for wheelwrights' work. Oak was grown as small and large timber for many building purposes. Many of the hedgerow and field oaks in Worcestershire nowadays are rather poor-looking specimens, often stag-headed. The die-back of oak branches at the top of a tree is not evidence of imminent death and such trees may change little in appearance for decades. Eventually dead branches fall and others grow and, although fungi may gain entry through holes and rot the heartwood (which is 'non-living' timber), the tree remains very much alive with its root-to-leaf conducting systems intact in the outer shell. The rotting wood and holey trunk of old trees provides house and home to innumerable insects, fungi and birds and they should, if possible, not be cut down. Such Worcestershire oaks are probably only about 100 years old — they can easily live for another 300 years.

Ancient and striking trees were often noted features in boundaries. Some of these arose from repeated pollarding, a practice which turns oak and ash into broad, gnarled trunks with a massive crown, from which branches were cut when the tree was pollarded. Some of these trees are extremely old. They are interesting historical features carrying a diverse

LEFT: Hollowfields Farm shows the hedge pattern of ancient countryside before clearance. (GHG)
RIGHT: Hollowfields Farm after conversion to a barley and wheat prairie. (GHG)

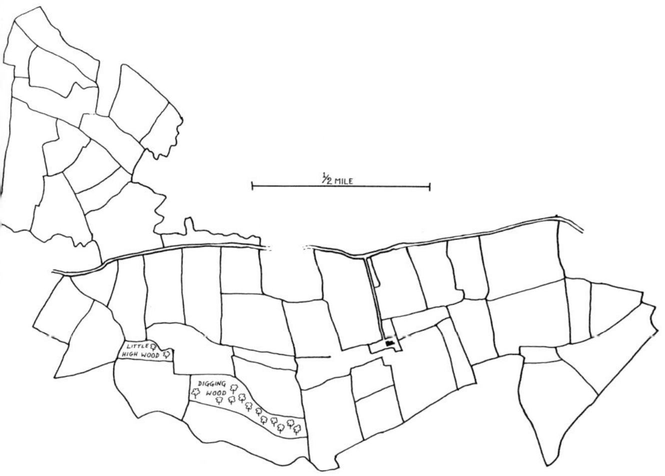

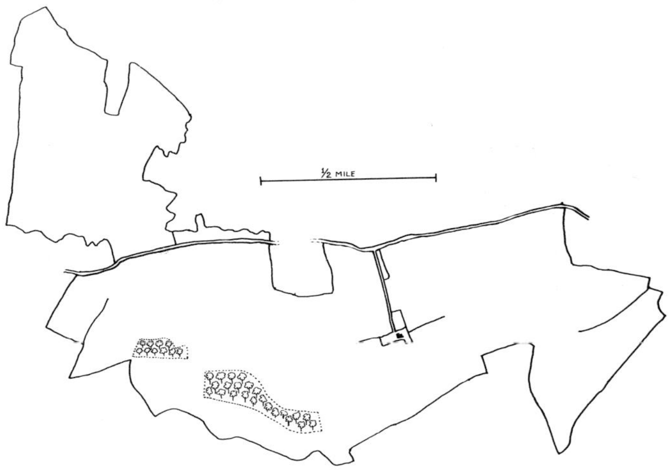

insect fauna and a large array of lichens and mosses, some rare, which in turn provide food and shelter for specialist invertebrates.

Although hedges are full of potential trees, flail-cutting regimes prevent them growing. This is partly because many farmers do not want their crops shaded by trees, or tall hedges, which would reduce productivity, and encourage some weed species. Trees and growing saplings are also a nuisance to flail operators — they get in the way and slow down work. There is no incentive to grow hedgerow trees and they are likely to decrease unless present attitudes change. Worcestershire's farmland could easily be almost tree-less in fifty years' time. There is no need to *plant* hedgerow trees — there are plenty of potential trees with good root-systems already in hedges which would soon produce trees if they were permitted to grow. Tree-planting beside hedges and on road verges is frequently doomed to failure as the planting site is often dry, the tree unprotected and usually neglected.

There has been a great decrease in hedges and hedgerow trees in the last fifty years. A random survey of fifty one-kilometre squares, comparing present-day hedges and traces of hedges with boundaries on old OS maps, showed that between 1920 and 1984, 21% of hedges had been removed. Extrapolation to cover the whole county estimates a reduction in Worcestershire's hedges from 12,100 miles to 9,550 miles, a loss of 2,550 miles. 55% or 5,300 miles of hedge were no longer stock-proof.

In a long period of agricultural prosperity and advance there has been great pressure on farmers to intensify cultivation through government monetary incentives, and by use of new strains of productive cereal crops, pest control chemicals, fertilizers and big, powerful machinery. Hedges and round small fields obstruct big machines and occupy cultivable land, so many hedges and their trees have been grubbed up to make huge fields. In one extreme example in 1982, at Hollowfields Farm near Hanbury, 600 acres were purchased by a developer and amalgamated into one farm. Public uproar followed as machinery more often seen in motorway construction moved in and uprooted 7.5 miles of hedge, many big trees and scrub. Fifty small fields were converted into three! Many hedges have been removed in other areas of the county but their removal has usually proceeded rather more slowly to create 60-100 acre fields.

Hedges provide much of natural history interest and importance in the countryside and they may act as corridors along which wildlife can traverse farmland. Trees and shrubs provide home for many insects, birds and mammals which use them as breeding, feeding

LEFT: Home Farm, Abberton, a large farm with big fields but many hedges retained. Circles show territories of reed buntings nesting in farmland. (GHG) CENTRE: Yellowhammer territories associated with the hedges. (GHG) and RIGHT: Chaffinch territories which are always associated with hedge trees. (GHG)

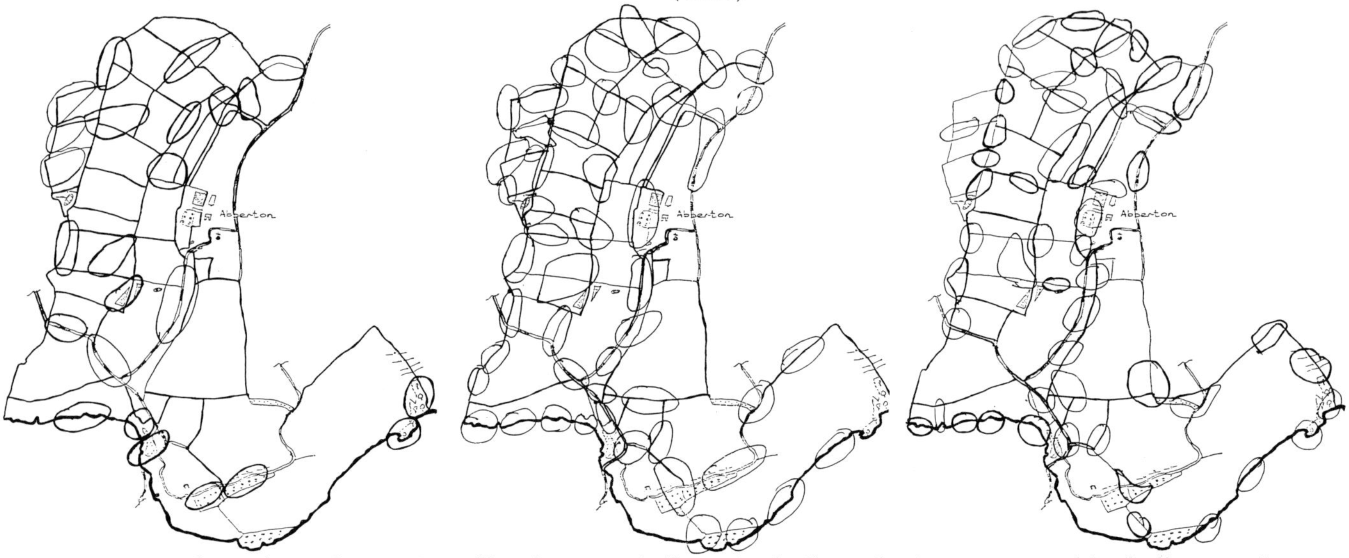

and resting places. Woodland ground flora including dog's mercury, bluebells, wood anemone and primroses, maintained on farms in hedge bottoms. Hedges are linear scrub used by both scrub and woodland birds. A common bird census on 600 acre Home Farm at Abberton, which has about six miles of hedge, showed 30 species of breeding birds in about 400 breeding territories associated with hedges and trees. The commonest hedge birds were chaffinches, yellowhammers, blackbirds, wrens and dunnocks, comprising about 40% of the total bird numbers found. Chaffinches were all associated with hedge trees, which they use as song posts, sources of caterpillars on which to feed young and sometimes for nest sites. Yellowhammers are scrubland birds, nesting in hedgerow bottoms, feeding their young on insects and later in the season feeding on weed seeds. If Home Farm, a well-run productive farm, had been treated like Hollowfields Farm, many bird species and about half the territories would have been lost, roughly 450 breeding birds. Counting the young they might have produced in a year (say two each), perhaps 2,250 individual birds would have gone. These rough calculations give some idea of the enormous changes in bird populations which have taken place on farmland in recent years. A rough estimate suggests a loss of about one million birds from Worcestershire due to hedgerow clearance and management.

The management and structure of a hedge, and the adjacent verges or headlands in cultivated fields greatly affect its natural history interest. Flail cutting soon knocks hedges into a nice trimmed shape with a square top which looks tidy but can be bad for a hedge and poor for its wildlife. It discourages bottom growth which laying once achieved. Add to this, over-enthusiastic flailing in spring (when birds are breeding and flowers opening) and similar treatment at harvest, it can be appreciated that many hedges are under severe stress. Heavily flailed hedges soon become gappy at the bottom and can evolve into a row of sticks with a few leaves on top. This is especially so if grazing sheep attack the hedge bottom. Rows of flat-topped low bushes with a post and wire fence on one side and no ground cover are a common sight. Hedges which have been laid survive reasonable flail treatment more successfully. If hedges are flailed less frequently, ideally every two to three years, the woody species have time to produce flowers and seed and the hedge will support many more birds, flowers and insects. It is often suggested that hedges trimmed into an A-shape give best results but such trimming may damage tops of shrubs and reduce flowering. A good diversity of height and irregularity in shape and form is probably more satisfactory for a wide range of wildlife.

White-letter hairstreak: the caterpillars feed on leaves of hedgerow elms. (WNCT)

There are, of course, big neglected hedges in the county which have formed massive scrubby tangles or, if livestock grazes beneath them, a row of trees, perhaps 20 feet tall. These big hedges often bear a profusion of flowers and fruit. For example, masses of white hawthorn flowers in spring are one of the joys of the county and the abundant berry crop they produce is an important food resource for resident thrushes (blackbirds, song thrushes, mistle thrushes), winter visiting thrushes (fieldfares, redwings), finches (hawfinch — the only finch that can crack haws), starlings and wood pigeons. Hedgerow berries of all sorts are important foods for birds. Big hedges also provide nest sites for larger birds, particularly magpies, wood pigeons, turtle doves, carrion crows, and occasionally herons. Lesser whitethroats are especially fond of tall, untrimmed hedges in low-lying agricultural land and grasshopper warblers sometimes breed in low hedges in the absence of more suitable habitat. A fifteen foot tall, thick hawthorn hedge near Little Comberton supported a thriving colony of breeding house sparrows for many years. As almost everybody knows, house sparrows usually build untidy nests of straw and feathers in holes, most often in buildings, but this colony built many nests of straw balls set in the thick hedge.

One central Worcestershire hedgerow site has the county's only colony of brown hairsteak butterflies, a beautiful but elusive insect. Adults emerge in mid-August and lay their eggs on new blackthorn growth. The white eggs overwinter between the thorns and the main stem and are thus vulnerable to flail-mowing. Agreement between the landowner and members of the West Midlands Branch of the British Butterfly Conservation Society has minimised disturbance at the critical period, an excellent example of co-operative conservation. Its relative, the white-letter hairstreak, is also hedge-dependent, feeding on elm leaves on farmland and at woodland edges.

Hedges with adjacent roadside ditches and verges are often excellent wildlife habitats, allowing many species to survive in an otherwise hostile environment. The hedge, verge and ditch can create in a few yards a miniature woodland edge, scrubland, marshland and grassland, depending on management. Road verges contain some scarce plants; however, rich verges are infrequent and require protection. About 60 sites, with a total length of about 15 miles and scattered throughout the county, have been designated as roadside nature reserves by WNCT, working with the County Council. The roadside reserves are marked with red posts and a schedule is kept by the County Council Highways Department and WNCT. Specific management of each reserve to enhance its interest would be ideal but is usually impractical because of insufficient resources. The verge cutters recognise the red posts and either do not cut the reserve or simply trim the edge. A danger is that untrimmed verges may be invaded by scrub which will shade out interesting flowering plants. WNCT will be pleased to hear of interesting verges, which will be surveyed and may be added to the schedule.

Floristically the most diverse road verges are usually on poor soil, often in limestone areas. If subsoils exposed by roadworks are left they are gradually invaded by a wide range of plants adapted to open conditions, without vigorous competition from the plants which flourish on fertile soils. The policy of spreading top-soil after road engineering and seeding with grass leads to a simple verge flora of little interest and attractiveness. Furthermore, flail-mowing adds yet more nutrients from finely chopped vegetation, and this leads to the most common of the county's road verge community: cow parsley and tall false oat grass. Both are attractive plants, providing the verge mowers give road-users a chance to enjoy

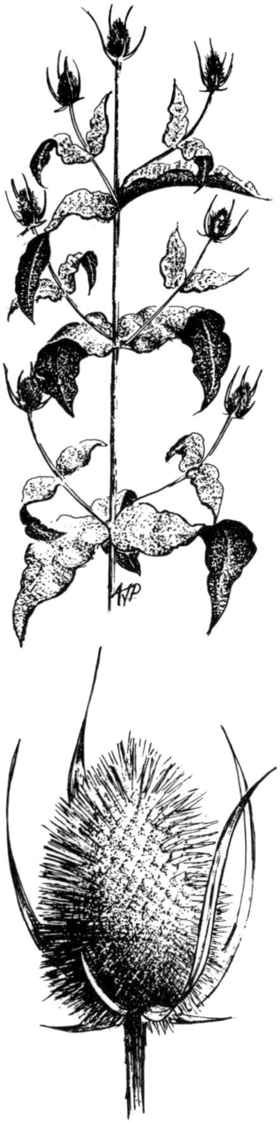

Roadside teasel. (WNCT)

them before they are cut down. A may-covered hawthorn hedge backing great banks of lacey white-flowered cow parsley in May is a sight for sore eyes, but sadly the cow parsley is often minced by a mower as soon as it flowers. The tall drooping stems of false oat last well into the autumn and are often festooned with spiders' webs, revealed by early morning dew — another pleasure for the commuting road-user. Hogweed is also common on slightly damper, fertile verges and is another plant whose dried stem and flowerhead withstands the weather into the winter and is used by spiders for web sites. Hogweed also attracts many insects to its flowers and older textbooks on insects show pictures of them covered with flies, plant bugs, and beetles. It is difficult to find such rich insect life now, on main road hogweed flowers which are next to intensive agriculture.

Roadside umbelliferous plants are mostly big plants and the practised eye of a botanically-minded driver can follow a succession of flowering through the spring and summer. In April and May there is the abundant cow parsley, followed by rough chervil in the hedge bottoms and swathes of the scarcer, shorter, ground elder in May and June. Hogweed, towering above other vegetation, starts to flower in June and the flowers of upright hedge parsley replace rought chervil in the hedge bottoms. In June and July some dry verges are carpeted with wild carrot, first with a convex white umbel (with one dark flower near the centre) which grows into an in-turned tight mass of hooked fruits. In June and July massive lacy stands of hemlock appear in wetter places and in late summer through to September great globular heads of wild angelica appear. On drier, often poorer soils, tall banks of wild parsnip, our only big yellow-flowered umbel, appear in July and August and at this time of year you may see the tall spires of small umbels of stone parsley. From mid-July, on limey thin soils, burnet saxifrage shows white flowers on wirey mid-green stems and it may be accompanied by yellowish-flowered pepper saxifrage with stiff dark green stems and leaves. Finally, for Worcestershire roadside umbel-spotters, look out for greater burnet saxifrage, a tall faintly cream, flowered plant which grows in roadside hedges especially in the Huddington-Himbleton-Crowle area, flowering from mid-July. Oliver Rackham thinks this plant may be associated with ancient woodland edges, surviving in hedges when the woods have gone. He calls it a 'circumboscal' plant! Interestingly, it flourishes in an area of Worcestershire's ancient countryside where there are still many woods, once part of the great Forest of Feckenham.

Road verges on poorer soils often support a good grassland flora and fauna with bird's-foot trefoil, clovers, medicks, ox-eye daisies, knapweeds, and a wide variety of meadowland grasses. These plant communities are good for insects and common butterflies (and their caterpillar food plants) such as meadow brown (smooth meadow grass), gatekeeper (bents, meadow grasses, fescues, couch), common blue (bird's-foot trefoil), large skipper (cock's-foot grass). Colonies of marbled whites also occur but these are usually restricted in distribution and require the presence of red fescue grass for the caterpillars. Roadside grasslands are also one of the few places where grasshoppers may still be abundant and spiders common. If ditches are present, marshland plants like meadow sweet, great hairy willow herb, and figworts may be there.

In the west of the county, where there are well-drained acid soils, different plant communities grow. Bracken is common and wood sage and betony occur. At Hartlebury, one such verge holds green-flowered helleborine, at one of its three county sites.

The variety and richness of the roadside reserves is easily demonstrated by a few examples. Near Gadbury Bank in the south-west of the county a verge on limestone soil has 80 plant species including pyramidal orchid, wild liquorice (milk vetch), wild basil, cowslips, yellow oat grass, upright brome grass, crosswort, lady's bedstraw, bush and common vetches, an hoary plaintain and is backed by a hedge of spindle, dogwood, hazel, hawthorn, oak, ash, wayfaring tree, bramble and spurge laurel with woodland plant herb Paris growing in the base. At Hazelor Bank, on the main A44 near Evesham, on poor limestone soil, a rich verge contains pyramidal orchid, bee-orchid, great butterfly orchid, spotted orchid, hairy violet, cowslips, chicory and sulphur cinquefoil, a naturalised garden escape. Bee-orchids are quite often found on disturbed, well-drained limestone-based verges in the south of the county. Near Romsley there is a verge with eighty species, including a large stand of wood horsetail (an uncommon plant), great horsetail, foxgloves, town-hall clock, and a hedge of hazel, holly, elder, oak and mountain ash. Near Ribbesford is a verge with lily-of-the-valley; at Great Witley a stand of butterbur and at Hanley Swan a remarkable verge of autumn crocus, *Crocus speciosa*, now long established but originally a garden escape.

Meadow crane's-bill, a straggling plant with attractive blue flowers nearly an inch across, commonly grows among rank grass and other plants on verges and in roadside hedge bottoms, which nowadays are its main habitat. At Broadwas, a road verge supports a flourishing colony of dwarf elder.

Bank voles, field voles, wood mice and uncommon yellow-necked mice all live in Worcestershire's hedges and verges, between them eating a wide range of stems, leaves, fruits and insects. These small mammals are preyed upon by kestrels, which are often seen hovering and hunting along road verges. The development of wide verges along the M5 motorway with dense, grassy, unmown banks, has provided an extensive field vole habitat and the hunting kestrel has become a common motorway bird. The larger mammals, foxes and badgers, use hedges on banks and in hilly areas as safe places to excavate dens and setts. Rabbits are abundant in some areas with warrens in hedge bottoms. In pre-myxomatosis days, when rabbits were an extremely common pest, hedgerows contained many warrens.

Hedge banks and bottoms are hibernating places for frogs, toads and newts, which crawl deeply in holes and hollows where they are protected from temperature extremes. Common lizards and slow worms are sometimes seen and, in the Wyre Forest area, adders occur and grass snakes may hunt along hedge bottoms for mice.

Hedges are good refuges for small rodents: LEFT: wood mouse, (RMB)
CENTRE: bank vole and (HB) RIGHT: field vole. (HB)

ABOVE: Hedge-laying. (DMG) BELOW: An unusual clump of horizontal old man's beard — a hedge climber, here with a mass of seed heads. (DMG)

ABOVE: White bryony and cleavers — hedgerow climbers. (DMG)
BELOW: Roadside tree-planting is now a welcome feature of new developments. (DMG)

Grasslands

In 1937 the land utilization survey of the county recorded 250,383 acres of permanent pasture, or slightly over 70% of the county's agricultural land. These pastures were on high ground with thin soils and steep slopes difficult to cultivate; on land liable to flooding bordering the main rivers, and on low-lying land such as Longdon Marsh and Defford Common, and on heavy clay soils (Coal Measures clays, Keuper Marls and Lias clays) especially the heavy glacial drift over Keuper Marl in the Wythall area. They were also classified as thin upland grasslands of the north and west, mainly suitable for sheep but easily burning out on sandy soils in summer; alluvial grasslands in the valleys of Severn, Avon and Teme where winter flooding was advantageous but spring and summer floods disastrous, destroying hay crops, and lowland grasslands situated where drainage was difficult, often on medieval ridge and furrow. Further south the Lower Lias clays were down to grass because cultivation with horses was extremely difficult: the heavy clays were called 'four-horse' and 'six-horse' land according to the number of horses need to pull a plough.

Over half the grasslands were then classified as 'ordinary Agrostis [bent-grassed] grasslands' with common bent as the most abundant species. Other better quality (in agricultural terms) grasslands contained rye-grass. Some contained too much rush and sedge and there were about 5,000 acres each of hill festuca grasslands and grasslands invaded by bracken, gorse and heather. Overall, Worcestershire's grasslands were agriculturally moderate to poor.

The Domesday survey showed that most meadowland was in the south of the county bordering the Avon and the lower Severn, parts of the Teme, along the Bow and Piddle Brooks and in the Longdon Marsh area. Meadows were more valuable than arable in Saxon, Norman and medieval times, providing fodder throughout the year for the oxen and horses which worked the land. Narrow meadows, cleared of woodland long ago, and lying between woodland and brook, can be seen on present-day maps. There are fourteenth century records of meadows being watered to encourage an early growth of grass in spring. Pitt, for example, writing in 1813, describes systems of channels and sluices designed to water

The wild daffodils of Duke of York Meadow in spring. (DMG)

both meadows and arable land from brooks. Near Kidderminster, contour line brooks were dug to carry water to previously barren, dry, sandy land, and one can be still be seen at Dunclent, where it gives the illusion of flowing uphill. Around 1810 there were about 100,000 acres of permanent pasture in the county and this rose to 184,753 acres by 1874 and to an all-time maximum of 260,585 acres in 1916.

Since World War II, management of lowland pastureland has changed enormously. Before then permanent pastures, although considered as 'Agrostis' or 'rye-grass' pastures, contained many other plants and were often enriched with deep-rooted nitrogen-fixing clovers and bird's-foot trefoil. Many meadows were rich in naturally occurring herbs and grasses. Some were plagued by the grass-parasite hay-rattle which depresses grass growth and productivity. The roots of many plant species growing together form a thick sward resistant to invasion by weeds. Breaking the sward and 'poaching' by over-stocking with farm animals allows unwanted creeping thistles and docks to become established. Without herbicides, they were difficult to eradicate, requiring repeating cutting before seed-set. The application of manure to pastureland encouraged the growth of nutritious grasses to the detriment of flowering plants, so improving agricultural productivity. As Pitt states 'it was better to put manure on greensward than tillage, if the land was sound and well drained'. Well-watered flood plain meadows gave a richer growth of grass than other meadows, benefiting from silt deposited by the floods, and were used as fattening pastures for stock.

During the last forty years many old lowland permanent pastures have been ploughed and re-seeded or used as arable land. Nowadays Worcestershire is still about half grassland and its grasslands may be categorised as short-term leys, improved permanent pasture and unimproved permanent pasture. Ecologically the differences between the three are great. Short-term leys are crops grown from seed and maintained for a few years: they contain few species of plants and have poor invertebrate populations. Improved permanent pastures are treated with fertilizers, and sometimes herbicides, to increase the growth of those grass species which fatten stock most quickly and to reduce the number of competing herbs and weeds. The application of fertilizer is easy, using a tractor and spreader, and the grasses which benefit from enrichment grow quickly and strongly and smother many herbs. Nettles, thistles and docks are controlled with herbicides.

Permanent unimproved pastures are not treated in this way although, traditionally, especially in lowland areas, they are treated with farmyard manure. Unimproved meadows are of the greatest natural history and conservation interest as they contain many species of grasses and herbs and large invertebrate populations, including spiders, craneflies, butterflies, beetles and their predators.

Biologically, Worcestershire's grasslands fall into three main categories, each with sub-divisions.

Limestone grasslands. One type occurs on the oolitic limestones on Bredon Hill and Broadway Hill: characteristic plants are carline thistle, wild thyme, hound's tongue, tor grass, meadow oat grass, salad burnet, rock-rose and several orchids. A second type grows on the Silurian Wenlock limestones in the west Worcestershire Hills: characteristic plants are the same, often with ploughman's spikenard and dyer's greenweed. A third type is found on the Rhaetic limestones of the ridge running north to south in the south-east part of the county. WNCT has three small reserves on the Rhaetic, at Windmill Hill near the Littletons, Cleeve Prior, Tunnel Hill near Wood Norton, and Badgers Hill on Bishampton Bank. There used to be extensive areas of limestone grassland in the Wood Norton-Sherriffs Lench district, now destroyed by cultivation and forestry. Characteristic plants are stemless thistle, bee orchid, pyramidal orchid, salad burnet, yellow-wort and thyme.

Acidic grasslands. These are found on volcanic rocks of the Malverns, Clent and Lickey Hills. Characteristic plants are mat grass, sheep's fescue, heath grass, tormentil, heath bedstraw and wavy hair-grass.

Neutral grasslands. There are three main types. One is found on the higher land of the Carboniferous coal measures or the Old Red Sandstone in the north and west, where drainage is poor. Characteristic plants are cowslips, adder's tongue fern, lady's mantle, bitter vetch, pignut and lousewort.

Another type clothes the flood plain meadows and water meadows of the Avon, Severn and the lower reaches of the Teme, often lying within great bends of the rivers and called 'ham' meadows. They have been important permanent pastures in Worcestershire since at least Domesday. Many were held as common land with complex systems of ownership, hay and grazing rights, with the plots marked by large stones. Buchanan gives a map of the common ham meadows near Powick and Gaut quotes a document dated 1508 dealing with disputes over encroachment onto these important common meadows. In more modern times fertilizing floods have become less welcome. As long ago as 1881 a great flood bank was built round the Fladbury ham: 4,647 yards long, fourteen feet three inches at the base, three feet at the top and four-and-a-half feet high, fitted with flood valves and sluices, with a drainage scheme using 10,000 yards of three-inch pipes behind it. The remains of this scheme can still be seen, but it no longer operates and the land is now regularly flooded. In more recent times the Water Authority and agriculturalists considered schemes to control the Avon floods more vigorously, so enabling the hams to be cultivated, but the cost was huge and the economic return low, especially in the face of European food surpluses, and these plans have been shelved. Nevertheless, some farmers have taken the risk and ploughed old ham pastures, but the resulting crops have often been flooded and spoiled. Nowadays the pasture of most ham meadows is improved and drained with consequent loss of species, but most are still there.

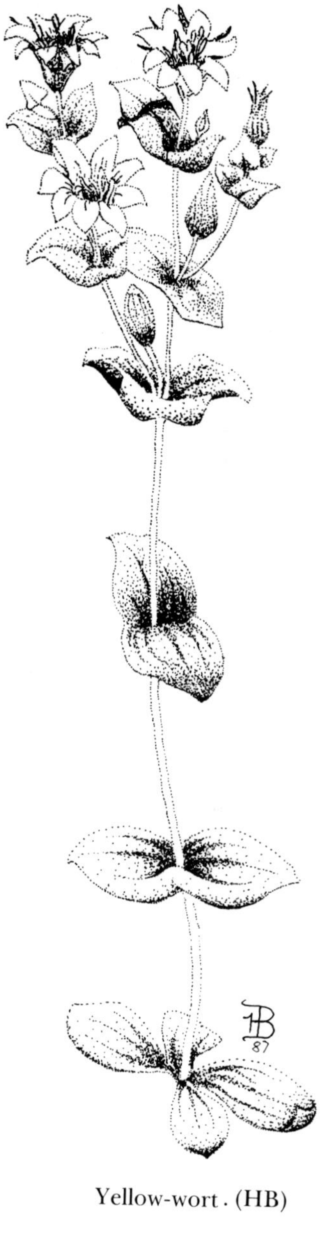
Yellow-wort. (HB)

Good examples of the original sward do exist, especially at Upton-on-Severn Ham, where Adrian Darby, a farmer and dedicated conservationist, has acquired a large part of the ham to manage in the tradition of hay-making and aftermath grazing. The distinctive plants of the Upton Ham are meadow foxtail, red fescue, meadow vetchling, pepper saxifrage, great burnet, adder's tongue fern, meadow saffron, common meadow-rue, bistort, meadow crane's-bill, parsley water dropwort, narrow leaved water dropwort and mouse-tail (on the disturbed soil at gateways). These plants occur on a few other unimproved ham meadows such as Twyning Great Meadow, straddling the county boundary near Bredon and at Uckinghall Meadow near Ripple. Dropwort, great burnet and the water-dropworts are characteristic plants of these meadows and in the past meadow saffron was common. Lees, writing in 1867, states that it was found 'in meadows all through the Vale of Severn from Bewdley to Tewkesbury. Too abundant' — can meadow saffron ever be too abundant? Undrained wet patches support marsh plants such as ragged robin, cuckoo flower and meadow sweet.

The Upton Ham and Twyning Great Meadow attract large numbers of birds in winter, especially after flooding. Huge flocks of gulls, lapwings, and starlings feed and roost, joined by up to

2,000 wintering golden plover and hundreds of dunlins. Bewick swans are often seen at Twyning in winter. Breeding birds include redshank, curlew, snipe and lapwing. This is an important site nowadays as inland wet meadow breeding sites for waders have become extremely scarce. The once numerous Avon and Severn valley breeding curlew, redshank, snipe and lapwing have greatly decreased during the last twenty years, due to increased land drainage, disappearance of marshy areas, the increase in silage harvested early in summer, and multiple hay-cropping, which destroys nests. As recently as the 1970s corncrakes still rasped and nested on the Avon ham meadows, but they have not been recorded for years.

The third type of neutral grassland is the *lowland grassland* permanent pasture and hay meadows found on the heavy clays and marls of the Lower Lias and Triassic in the south-west and east of the county. Most of these are on ancient ridge and furrow fields difficult to cultivate and enclosed in the seventeenth century or by later Parliamentary Enclosure Acts. These fields were numerous and ridge and furrow, 'fossilised' beneath permanent pasture, is a striking feature of aerial photographs taken in the late 1940s. Since then most have been ploughed and under-drained. Modern, powerful tractors and ploughs can break down ancient ridge and furrow and fertilizers enable the rather poor land to be cropped. Some such fields remain but most of these are now improved pastureland with a limited number of plants. The characteristic plants of unimproved lowland meadows are green-winged orchid, meadow saffron, cowslip, dyer's greenweed, saw-wort, adder's-tongue fern and yellow (or hay) rattle.

Despite the losses, Worcestershire is an important county for unimproved lowland meadow pastures and efforts are being made by both WNCT and NCC to conserve good examples. Some of the best are now SSSIs and nature reserves. WNCT owns a nationally important site at Fosters Green Farm, near Hanbury. Until disafforestation in 1591, this land lay in the Royal Forest of Feckenham and it is quite likely that is was wooded until then. The small farm may have been cut out of the woods and this contention is supported by the species-rich hedges and the persistence of wood anemone, violets, goldilocks, primroses, bluebells and dog's mercury and other woodland plants in the fields. The earliest historical reference is from 1640 when Fosters Green was a small farm with meadow, pasture and two small orchards. It was mapped in 1772 when the land passed to a Mr Eades, whose name is now attached to the large hay meadow. By 1840 several hedges had been removed and by 1927 Eades Meadow was one big field of 14 acres with a stand of fruit trees. Nowadays the big oak trees in the meadow mark the old hedgelines. The record shows that the eastern end of the meadow may never have been ploughed, since it was cut out of woodland. Eades Meadow has been managed without chemicals or soil disturbance for a long time and the sward is rich in plant species. On average a square metre contains around 37 species of grasses and herbs. An improved pasture may contain up to ten at most.

Meadow saffron, the 'true' autumn crocus, *Colchicum autumnale*, grows in Eades Meadow, which is all that now remains of the once-extensive *colchicum* meadows of the district. This curious plant is also called naked

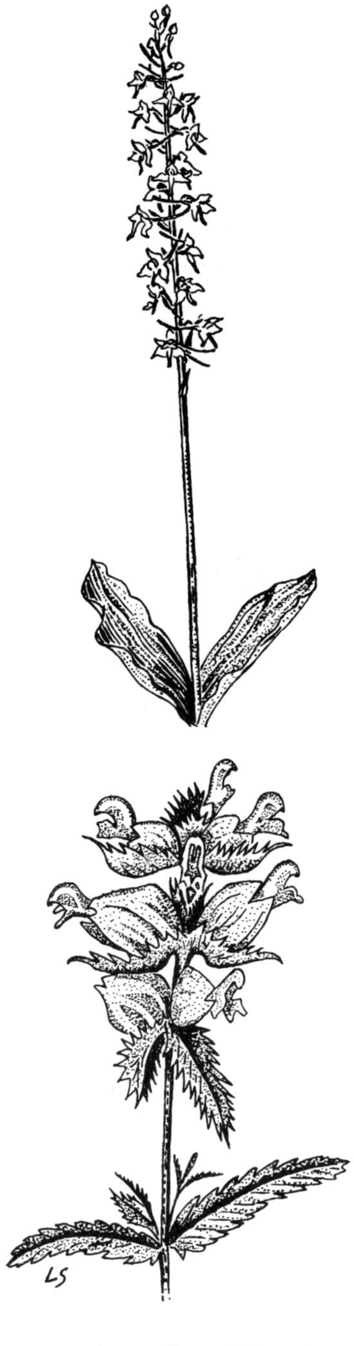

Great butterfly orchid and yellow rattle. (LS)

ladies because its crocus-like flower appears in September without leaves. The flower dies down and, in the following spring, broad, strap-like leaves and seed-pods are produced. These are poisonous to cattle when green but harmless when cut and dried in hay. The seeds and corm were once used to produce *tinctura colchici* for the treatment of gout. The number of meadow saffron plants in Eades Meadow has decreased in recent years, with no

Meadow flowers. (WM)

proven explanation, but perhaps because of the loss of many shading hedgerow elms from elm disease, and dry summer weather. The plant also grows in woodland and along rides and may require a fairly moist, shady environment.

Another interesting plant of old pastures (and woodlands) is adder's-tongue fern. The spore-producing generation is small but noticeable as a pointed, oval, upstanding leaf and a vertical spore-bearing spike alongside it. The fern spreads underground for long distances. The plant is long-lived and lives in association with mycorrhizal (ie: root-linked) fungus which enables the fern's roots to obtain nutrients from rather damp neutral soil formed by the breakdown of plant material. The small sexual generation of the fern lives in similar soil conditions, which are found in stable, old, undisturbed swards of permanent unimproved pastures. Ploughing and re-seeding quickly destroys it. Moonwort, a rare heathland plant in Worcestershire, has a similar life history and is found in pasture on drained sandy soils.

130 grassland and woodland plant species have been found in Eades Meadow, ranging from large oaks to small grasses, including cowslips, lady's smock, ragged robin, buttercups, knapweeds, violets, bugle, spotted orchids, green-winged orchids, marsh marigolds, twayblades, adder's-tongue fern, pepper saxifrage, ox-eye daisies, quaking grass and twenty-four other grass species.

Perhaps the most amazing thing about Eades Meadow to the passing observer is that it never looks the same for two days running during the spring and summer. Every time you look over the hedge the colour has changed, depending on which flowers are in bloom, the colour of the ripening grasses, which seed-heads are ripe, and the flickering interchange of texture and colour caused by the passing breeze. In spring there is a purple haze of green-winged orchids against a yellow background of buttercups and cowslips. In summer the knapweed flowers glow against the yellows and browns of ripening grasses. In autumn the delicate lilac of the meadow saffron shows softly against the green of new-growing grass. When you kneel in the meadow and part the plants you are amazed at the variety of leaf and flower shape and the abundance of insects and spiders. Eades Meadow is a remarkable place. Its rich diversity is the product of a particular style of farming. Grazing and hay-making are essential. Without them the meadow would be invaded by coarse grasses and scrub, and tough, dominant plants would swamp the herbs. Some fertilization by farmyard manure is also necessary to put back fertility removed by the cattle and hay. Eades Meadow is part of a working farm and it can only be visited on special open days organised by WNCT, but much can be seen by looking over the roadside gate.

Eades Meadow is the epitome of Worcestershire's lowland meadows and few others are as remarkable, although many contain different mixtures of plants. The WNCT's Duke of York Meadow, next to a pub of that name, at Berrow, is a wild daffodil meadow, and an example of a meadow type once-common on the Worcestershire-Herefordshire border. It is also a hay meadow with many other plants, including great burnet, cowslips and green-winged orchids. At Brotheridge Green there are five small hay meadows of several types: grassy fields with quaking grass, sweet vernal grass, crested dog's-tail, and tall fescue; herb-rich meadows with yellow rattle and lesser knapweed, and sedge-dominated wet fields with abundant carnation sedge and meadow sweet. One has an astonishingly dense display of green-winged orchids in spring, and adder's-tongue fern is found. WNCT also leases several semi-improved adjacent fields and plans to 'un-improve' them by using traditional techniques and applying seed or hay from the unimproved fields.

Near Kempsey there is a group of fields containing 115 grassland plants including saw-wort, dyer's greenweed, wood small-reed (usually found in woods), brown sedge (scarce in the county), sneezewort and many others. WNCT owns a meadow at Poolhay where there are masses of green-winged orchids and it is one of the few county sites for meadow thistle.

Before 1940 over 70% of Worcestershire was grassland, with an enormous number of species-rich unimproved pastures. Since then most have been changed and only a small

ABOVE: Duke of York Meadow, purchased by the WNCT for its splendid show of wild daffodils. (DH) LEFT to RIGHT: Green-winged orchids, indicators of ancient meadows, great butterfly orchid, meadow saffron at Eade's Meadow — one of the finest old meadows in England — and spiny restharrow. (All DH) BELOW: Small elephant hawk moth. (DH)

LEFT: Large skipper — grassland butterfly, (WNCT) CENTRE: meadow grasshopper (HB) and BELOW: meadow brown — a lowland grassland butterfly. (LS)

number remain, though perhaps more than in some parts of England. Modern farming methods, the decrease in stock, the increase in short-term leys and the vast increase in cereal arable farming have all taken their toll of old pastures. It seems important to maintain and conserve some of these old fields with old-style management.The ideal conservation regime for lowland meadows is to cut the hay late, between mid-June and mid-July after seed has set and fallen, leave about two months for a growth of new grass, then graze in autumn and even winter, taking care to avoid damage by 'poaching'. Grazing is stopped in late February and the hay crop is allowed to grow. An even spread dressing of about 8-10 tons of farmyard manure per acre is applied every two to three years in the autumn. Harrowing at the end of March to break up a mat of dead stems may be helpful and rolling may be needed if the ground is disturbed. For old pastures the best grazing animals are cattle or sheep. It is difficult to maintain rich meadows with horses, as they are selective feeders, taking only some species and often pulling them up. Large masses of horse droppings also create zones of exaggerated fertility. This is unfortunate because so many of the small fields selected for pony paddocks are often interesting pastures. With care, these can be maintained, and even improved (in the genuine sense of the word) with careful management, including hay-making and harrowing to spread the muck. The damage and destruction is obvious in the many horse paddocks converted to muddy patches or dock-fields by bad management.

Although it would probably take centuries to re-establish a pasture like Eades Meadow it is possible to create a fairly herb-rich pasture within a few years. WNCT's advisers are helping in a project of this type near the windmill at the Avoncroft Museum of Buildings, near Bromsgrove. Here the first steps were the removal of the fertile top-soil and control of weeds with herbicide. A suitable seed-mix was sown and we await events. Further weed control may be necessary. Meadow establishment is not easy and partly depends on good, properly collected and treated seed.

Despite the example of fields like Eades Meadow it is difficult for anyone who did not experience the flowery fields of the 1940s and earlier to understand the difference they made to the countryside. When a child of those days wandered into any meadow (and woebetide those who trampled the hay) he or she entered another world. At every step among the sweet-stemmed tall grasses, and flowers such as ox-eye daisies, buttercups and bitter-tasting sorrel, grasshoppers leaped and whirred, daddy-longlegs blundered into the air, innumerable small moths flew about, and meadow browns, ringlets, and a whole variety of beetles, bugs and other insects could be found. On peering into the depths spiders were seen everywhere, in thousands, preying on the abundant invertebrate life. Beyond a child's vision the turf was full of invertebrates, large fat grub-like crane fly larvae, predatory beetles and their larvae, springtails and tons of earthworms. And moles were a plague! Moles eat earthworms and many fields were covered with molehills which broke the sward and let in weeds. No wonder moles were trapped and hunted with such vigour. The village molecatcher was an important person in grassland maintenance. Old pastures were also important feeding places for many birds. Starlings, blackbirds and song thrushes raised their young on invertebrates dragged from the sward and wintering fieldfares and redwings attacked the same resource. In May and June flights of cockchafers (now a rarity) and grassland moths provided food for bats. Old pastures are some of the most remarkable of close-packed, species-rich biological communities and their loss has greatly impoverished Worcestershire's natural history.

Green-winged orchids in Eades Meadow. (GHG)

LEFT: Eades Meadow — an extremely rich grassland of national importance. (DMG) CENTRE: Common spotted orchid — found in lowland meadows. (AWC) RIGHT: Bee-orchid often — found on limey disturbed road verges or grasslands. (GHG) BELOW: Monkwood Green — ancient common grassland by Monkwood. SSSI for its plants and insects. (DMG)

ABOVE: Ringlet — grassland butterfly. (WNCT) LEFT: Mating burnet moths — a daytime moth of old grasslands. (DMG) RIGHT: A pair of common blue butterflies — found where bird's-foot trefoil abounds. (JRH) BELOW: Rough grassland with clumps of gorse, invading scrub and scattered trees — good for butterflies, tree pipits and linnets. (DMG)

The pools and marshes at Feckenham Wylde Moor reserve. (DMG)

The Wetlands

Worcestershire's low-lying, fairly level central plain, dominated by the rivers Severn and Avon, with many streams and narrow river valleys in the north and west, must once have held large areas of marshland. Much of this was drained during land improvement schemes in the eighteenth and nineteenth centuries. Pitt, writing in 1813, states that the Earl of Coventry at Croome was an especially skilful drainer — '. . . his part of the county was a morass not half a century back, and is, at the present time (though formerly on moorish soil) perfectly dry, sound for sheep and cattle'.

The last great marsh in Worcestershire was the Longdon-Pendock-Eldersfield Marsh complex which defied the drainers for many years. In the seventeenth century Longdon Marsh probably extended to over 10,000 acres and was the remains of the great tidal estuary of the Severn above Gloucester. The flat land of the whole area is on virtually the same level as the River Severn, between ten and thirteen metres above sea level, so making drainage difficult. The area was surveyed by the celebrated engineer Brindley in 1763, and Hall in 1788, for drainage, but nothing was done until 1861-1866, when drainage operations were carried out under the Land Drainage Act. A survey at that time calculated that the marsh received water from an area of 17,000 acres and that, when the Severn was in flood, its water level was higher than in the marsh. The drainage scheme broke down and the land was not drained until after enclosure in 1872. Flooding still remained a problem but controls gradually improved, especially after deep dredging of the drainage ditches and brooks by the Water Authority in the 1980s. Nevertheless, a huge flood covered the area in February 1990, attracting large numbers of wildfowl.

The effect of tides on the Severn was considerable, and high spring tides ran up to Upton-on-Severn, holding up freshwater in the river with a consquent rise of 18 inches at

Worcester. Brackish water reached far inland and plants associated with coastal conditions grew along the Severn and in parts of Longdon Marsh. The Severn was slowly improved for navigation in the nineteenth century, although tidal influences did not cease until locks and weirs were built at Tewkesbury in 1868, and below Gloucester in 1871. After that the low-lying riverside and marsh areas were inundated with freshwater and salt-associated plants disappeared, to be replaced by freshwater marshes.

In 1867 Lees was writing of the botanical excitements of Longdon Marsh and doubting the likelihood of success of the recent attempts at drainage. At that time the marsh soon flooded in wet weather to cover an area of 3,000 acres, having all the appearances of a large natural lake. He writes 'Even in summer a few days' continued rain inundates these flat meadows, and the mowers then have to cut the grass knee-deep in water', and 'in autumn the marsh is covered with geese'. It is apparent from other reports, mentioned by Harthan in 1947, that the area was rich in waterfowl, much visited by seabirds and waders, and contained extensive reedbeds with bearded tits, bitterns and marsh harriers, similar to undrained fens and saltmarshes of East Anglia — not the scenario we expect from modern Worcestershire.

Some of the outstanding records from Longdon Marsh were sea club-rush, marsh pea, brown sedge, meadow thistle, yellow loosestrife, narrow-leaved water dropwort, parsley water dropwort, great burnet, golden dock, flowering rush, water violet and bladderwort. Corky-fruited water dropwort was also frequent along the nearby Severn. Many of these plants are now absent or scarce in the county.

Jack Green, in his booklet on *Butterflies of Worcestershire*, writes that a small colony of swallowtail butterflies bred near Eldersfield (part of the Longdon Marsh complex) until about 1920. They were, perhaps, the last remaining descendants of the swallowtails that once flew over Longdon Marsh. Interestingly, milk parsley, the foodplant for the caterpillar of this fenland butterfly is not listed in county floras. Was is overlooked or did the swallowtail caterpillars eat something else?

A WNCT survey in 1977 showed that Worcestershire contained about 375 acres of marshland and nearly 400 acres of wet grasslands. In 200-300 years, vast areas of marshland have been drained. Most of the remaining fragments are found in the flood plains of the Avon, Severn and Stour, the most important being in the Stour valley, near Kidderminster, at Wilden, Stourvale, Puxton and Caunsall. The first two are SSSIs and there is a third marshland SSSI at nearby Blake Marsh. These sites have rich and varied flora, with county strongholds for southern marsh orchids and marsh cinquefoil. The Stour marshes have developed on the narrow alluvial flood plain overlying Triassic sandstones, an important aquifer which has many springs in the valley. Past river engineering raised the river bed and this and the nearby elevated canal have helped raise the water table to the surface of the soil, providing conditions in which marshes have developed. Although the present Stour marshes have arisen from human activity they are of great conservation and natural history interest. They are at risk for several different reasons: first, through lack of grazing by cattle, which prevents the marsh being invaded by willow and alder scrub in the process of normal ecological succession, leading to wet carr woodland; second, drainage of nearby land for agricultural improvement may lead to drying out and loss of marshland habitat; third, river and canal engineering could drain the marshes. Unfortunately, flood alleviation works, funded by the Ministry of Agriculture and Fisheries, on the river adjacent to Wilden Marsh SSSI, a WNCT reserve, caused great damage to the marsh, with loss of areas of marshland flora. The bed of the river was considerably lowered by work completed in 1979 and it now runs in a much deeper bed below a band of clay running under the marsh. Prior to this work, water from the nearby hillside sandstone aquifer fed into the marsh, keeping it wet

Snipe — a typical marshland bird. (BW)

throughout the year. After the work, the water followed an easier course through gravel beds beneath the clay, so leaving much of the marsh high and dry. Sadly, before the work was done by the Water Authority, WNCT received the engineer's assurance that the marsh would not be damaged. Clearly, the implications of the scheme had not been understood and, after years of legal wrangling, WNCT received some monetary compensation for the damage to the SSSI, although the Authority would not formally admit negligence.

Plants which have declined in various parts of Wilden Marsh after the flood alleviation scheme are black sedge, wood club-rush (which requires thoroughly wet conditions), yellow flag, reed canary grass, marsh cinquefoil, cyperus sedges (a pond edge species), southern marsh orchids, lesser valerian, marsh woundwort, sneezewort and marsh pennywort. Wetland birds also declined: snipe from fifteen pairs to one pair, redshank and yellow wagtails from three pairs to none. The marsh was also an important wintering area for snipe, with up to 200 present, now reduced to about ten birds. Despite all these declines, Wilden Marsh remains an important marshland and other species include marsh marigold, tubular water-dropwort and lesser water-parsnip which grows in the ditches. The reserve contains a mixture of habitats, including dry unimproved pasture which is grazed, tall herbaceous vegetation with meadow sweet and great hairy willow herb, marsh and swamp, pools and ditches, willow scrub and patches of willow and alder woodland. A wide variety of birds breed, including sedge warblers and reed buntings. Settling pools associated with the nearby sugar-beet processing factory attract green sandpipers, common sandpipers and little ringed plovers. Marsh warblers and Cetti's warblers have occurred and there are also wintering water pipits on the lagoons.

Two other important marshes lie on the north side of Kidderminster: Puxton and Stourvale lie at the heart of the old carpet manufacturing district and are almost encircled by housing estates. They are used as rough grazing land and are now designated 'open space' so are likely to be maintained. Earlier tipping has reduced the size of Puxton. Like Wilden they are criss-crossed by drainage ditches. Flora is much the same as at Wilden and in both cases just over 100 flowering plant species have been recorded. Caunsall Marsh, between the Stour and canal at Cookley, just north of Kidderminster, has been almost completely invaded by alder woodland and has a limited flora.

Several of the brooks to the east of Kidderminster have been dammed to form mill-pools, particularly near Blakedown. These are now often filled with silt and show a succession of habitats, from open water to reed beds, beds of emergent vegetation, swamps, marshes, wet scrubland, willow and alder woods. Marshes have also developed round pools and lakes in other parts of the county with good examples at Westwood Great Pool, Lower Bittell reservoir, Ombersley Park, Cofton and Stanklyn.

Marshland sites are scarce along the Severn and the most important are associated with old pits from which clay was excavated to make bricks and tiles. There are thirteen sites; all have been long disused and many have developed into wet woodlands with large stands of willow and alder, sometimes over water. One of these now holds the county's main heronry, with about twenty breeding pairs in some years. Where there are open pools in the pits they are frequented by breeding mallards, moorhens and coots and a variety of wintering ducks. Where grazing has occurred and prevented invasion by scrub and trees, there are interesting flora.

Some of the most interesting of brick-pit wetlands and marshes have developed north of Worcester and some are designated as SSSIs. At Grimley there are two sites with varied habitats including marsh, swamp, tall herbaceous vegetation on both wet and dry ground, willow wood and scrub, open water pools and the adjacent river Severn. This is an extremely rich site with nearly 200 species of flowering plants, including marsh stitchwort and frogbit, over 100 birds, twelve mammals, frogs, newts, grass snakes, an abundant freshwater invertebrate community and eleven species of dragonfly, including the uncommon red-eyed damzelfly. Northwick Marsh, on the opposite side of the Severn, is more open and in winter is an important roosting and feeding place for birds such as snipe, jack snipe and meadow pipits. Invasion by Himalayan balsam is a worrying threat to these excellent wetland sites and already great ten-feet high thickets of the stuff grow in some places, smothering all other vegetation. Marshlands are also found in most gravel workings, which are extensive north of Worcester at Holt and Grimley.

There are some small sites and several large marshland sites on flat ground, usually at the bottom of drainage slopes, fed by springs and small streams. Two of the most important are Ipsley Alders, in Redditch, and Feckenham Wylde Moor. Both are base-rich marshes. Ipsley is based on a layer of sedge peat about five feet thick, lying on Keuper Marl clay. It receives spring water which originates from hills on the Warwickshire border to the east, which carries a high concentration of calcium sulphate. The predominant marshland plants are rushes and sedges, including dense stands of blunt-flowered rush, a typical fen species scarce in Worcestershire. Other notable flowering plants are hemp agrimony, spotted orchid, marsh and fen bedstraws, and the usual waterside plants. Most unusual is marsh stitchwort, found at only a few sites in the county. Mallard, teal, moorhen and coot are often seen on the pools, which are one of the few county sites for the black darter dragonfly.

The ancient Feckenham Wylde Moor was an extensive marshland on peaty soils overlying clay. Most of the marsh was drained long ago, following enclosure in 1833. The present marsh, a WNCT reserve, contains an old drainage system which has largely failed and the meadows have reverted to a mixture of wetlands.

Excavations for a gas main across the site increased wetness, probably by further breaking the drainage system, and an extensive reed bed has developed on wet ground. Like Ipsley Alders, the marsh receives calcium sulphate-rich water from springs supplied from low hills to the north, and it has developed a similar base-rich marshy grassland and fen flora with large stands of lesser pond sedge, blunt-flowered rush, and soft rush with reedmace, marsh thistle, wild angelica, meadow sweet, cuckoo flower and ragged robin. An unusual plant is the sea club-rush, which grows at a few other sites in the county, usually associated with salt. Bogbean grows in small pools, and cowslips, yellow rattle, adder's tongue fern and common orchids in drier grassland. Alas, lesser water plantain, grass-of-Parnassus, common butterwort, great sedge, and bog-rush, all recorded by nineteenth century botanists at Feckenham, have not been found in recent times.

Since WNCT acquired the site in 1981, one large and a number of small pools have been excavated and have attracted an excellent dragonfly fauna of twelve species, including the black-tailed skimmer, four-spot libellula and emperor dragonfly. Moorhens, ruddy-duck and mallards have nested and the pond has been colonised by stoneworts.

Both Ipsley Alders and Feckenham Wylde Moor marshes are based on thick peat over clay. These peat deposits have both been examined for preserved pollen grains and plant remains. Parts of Castlemorton at Malvern are also marshland and this differs from the others in being fairly acidic.

Wet meadows are similar to marshes, with the water table near the surface for much of the year. They usually lie on heavy lias or Keuper marl clays and have been subject to under-draining, which has eventually silted and failed. These fields are usually subject to

LEFT: Sea club-rush at Feckenham — a plant sometimes associated with salt and (RMB) RIGHT: blunt-flowered rush — abundant in Feckenham and Isley Alders marshes. (RMB)

flooding and are grazed and perhaps cut for hay, which prevents invasion by scrub. The predominant plants are grasses and rushes — commonly tufted hair grass, marsh foxtail, hard rush, soft rush, jointed rush and herbaceous plants such as creeping jenny, devil's bit scabious, bog stitchwort, marsh bedstraw and ragged robin. A few sites, like Ashmoor Common, are richer, with great burnet, sneezewort, southern marsh orchid, marsh pennywort, marsh marigold and tubular water dropwort. Ashmoor is further enhanced by a pool with emergent vegetation, and 154 species of flowering plant have been recorded. Some of the notable species are many-stalked spike-rush, common spike-rush, nodding bur-marigold, narrow-leaved water-dropwort, parsley water-dropwort, and common spotted orchid. Modern intensive under-draining of agricultural land has led to the disappearance of many large and small marsh and wet meadow sites and those which survive are often on the old commons which were not enclosed, for example Ashmoor, Marsh, Smithmoor, and Castlemorton Common.

Brine pumping has now ceased in Droitwich itself and the river and canal are less salty than in the past; many changes occurred with the establishment of Droitwich New Town. However, several small saline pools exist at old salt works and there is natural upwelling of brine at the Flash Pools at WNCT's Upton Warren reserve. These sites all support halophytes (salt-loving plants) and there is a small saltmarsh bordering the Flash Pools. The latter increased in extent and salinity between 1975 and 1985, with changes in distribution of the main halophytes, sea marsh grass, sea spurry and hastate orache. Several salt-tolerant species of grass known to adapt to a wide range of saline conditions are also present: couch, creeping bent grass, fine-leaved red fescue, meadow barley and perennial rye grass.

In the past salt springs were quite common in the flat parts of central Worcestershire, with examples at Defford Common, Pirton Common and at Saldon near Himbleton. Lees visited the last in July 1856 and saw an oozy salt spring with reflexed saltmarsh grass, saltmarsh rush and Babbington's orache. Where are they nowadays? Nobody seems aware of their existence.

Acid bogs are common in the north and west of Britain but rare in Worcestershire, with one small but outstanding example at Hartlebury Common and a few tiny remnants on the Clent Hills, near Hagley, in the Wyre Forest, on the Malvern Hills and at Castlemorton Common. At Hartlebury the bog contains areas of bog mosses (sphagnum) and acidicolous

LEFT: Reed warblers nest in reed beds throughout the county. (WNCT) and RIGHT: grasshopper warblers are found in both wet and dry scrubland and rough grassland. (WNCT)

plants — sundew, cranberry, common cotton-grass, cross-leaved heath, marsh cinquefoil, marsh violet — various sedges and rushes, and grasses, particularly purple moor grass. The bog appears to be declining and drying out and is now dominated by common cotton grass in the centre and purple moor grass at the edges. The sundew, dependant on wet sphagnum, has greatly decreased. Hartlebury Common is a Local Nature Reserve and efforts are being made to re-envigorate the bog.

The common reed tends to form tall dense stands in water up to about one metre deep and in wet soil. It grows in a wide range of conditions, and can withstand some pollution, high levels of nutrients and is usually found in base-rich conditions, avoiding acid soils and bogs. Reeds have been widely used for thatching in the past, particularly in East Anglia. In Worcestershire the old-style thatch was usually of wheat straw and the modern reed-thatch now used on timber-framed houses is imported. Due to land drainage the large reed-beds once found at Longdon and in other low-lying areas have disappeared. There are no major stands of reeds in present-day Worcestershire; the largest have developed the fringes of ornamental lakes, reservoirs, subsidence pools, gravel pits and along canals, rivers and ditches. The most extensive are at Westwood Great Pool, Hewell, Turnhill and Thorngrove. Reed beds have been expanding at Abberton and Feckenham Wylde Moor in recent years. Narrow linear stands grow at the margins of the river Avon and its tributaries, along the Worcester-Birmingham canal, and in many ditches in the Avon Valley and at Longdon. The five-mile reed bed which used to fill the Salwarpe Canal has now been extensively cleared during canal renovation work.

Reed beds are often important sites for birds; reed warblers, bitterns, bearded tits and marsh harriers are almost entirely confined to reed beds during the breeding season. Of these, reed warblers are quite common at the bigger reed beds and along the linear bed bordering the river Avon and the Birmingham-Worcester canal, but the biggest of all Worcestershire's colonies, which occupied the Salwarpe Canal, has been largely destroyed. There are big colonies at Westwood, Hewel and Thorngrove. Reed warblers are summer immigrants, arriving in May. The nest is usually woven round several vertical reed stems two to three feet above water, and lined with parts of reed flower heads, pulled off by the birds with considerable difficulty, so much so that deserted nests are quickly demolished or robbed of building material. The birds spend much of their time within the reeds but early in the season seek insect food from nearby willows and other trees. Reed warblers are quite frequently parasitised by cuckoos and parasitisation reaches high levels in the small reed beds by the Avon, presumably because the female cuckoos can easily find the warblers' nests, where there is little cover.

Bearded tits are uncommon in Worcestershire, with virtually no records until the species erupted from its East Anglian and Dutch breeding grounds (where enormous populations had built up on the Dutch polders during reclamation of land from the sea) in the 1960s

RIGHT: Marshland scene — old pollards and hunting kingfisher, (BW) and BELOW: a grass snake hunts in ponds and grasslands for frogs, mice and insects. (BW)

and early 1970s, with a Dutch-ringed bird caught with others at Westwood Great Pool in 1966. Since then winter records have continued at various reed beds in the county and, although summer records have followed, there is no definite proof of breeding. Bearded tits feed mainly on reed seeds during the winter and use reed heads in nest building. Bitterns have not bred in Worcestershire since the disappearance of the huge Longdon Marsh reed beds but there are occasional winter records, often in strange places, such as the bird that spent a winter's morning sitting in a plum tree in a garden in Naunton Beauchamp! Marsh harriers are also unknown but, when we look over Feckenham Wylde Moor, we often dream . . .

Many small birds feed in reed beds during the summer on the abundant insect population. Aphids on reeds are an important source of food sedge and reed warblers and foraging warblers and tits of many species are common. Reed beds over water are also important roost sites during autumn migration for swallows and sand martins and, before population declines of the latter since 1978 and the former in 1969, big roosts could be found at Westwood Great Pool and in smaller reed beds. Sand martins from the Westwood roosts were later found breeding as far north as central Scotland and on migration in southern England and north Africa. Several Worcestershire ringed swallows have been found wintering in South Africa. Starlings also use reed beds for roosts, especially in late summer, when the reeds may be weighed down by local juvenile and adult birds. Yellow and pied wagtails, reed buntings and occasional corn buntings also roost in reed beds, and they are attractive to pheasants for whom they provide important cover.

A mixture of herbaceous plants form tall stands in some marsh and wetland areas, particularly where the water table is not at the surface all year round and there is no grazing. Such areas are found by rivers and streams and round pools and lakes, such as by the Avon; breeding marsh warblers use such stands in a small part of the county. The main plants are great hairy willow herb and meadow sweet, often mixed with tufted hair grass, nettles and great and lesser pond sedges. There are often associated willow trees and scrub. These sites are important breeding places for sedge warblers and reed buntings and foraging grounds for warblers and many common birds. In summer blackbirds and song thrushes frequently drop into the tall vegetation in search of food and the latter commonly collect snails, which they smash on an anvil on nearby drier ground.

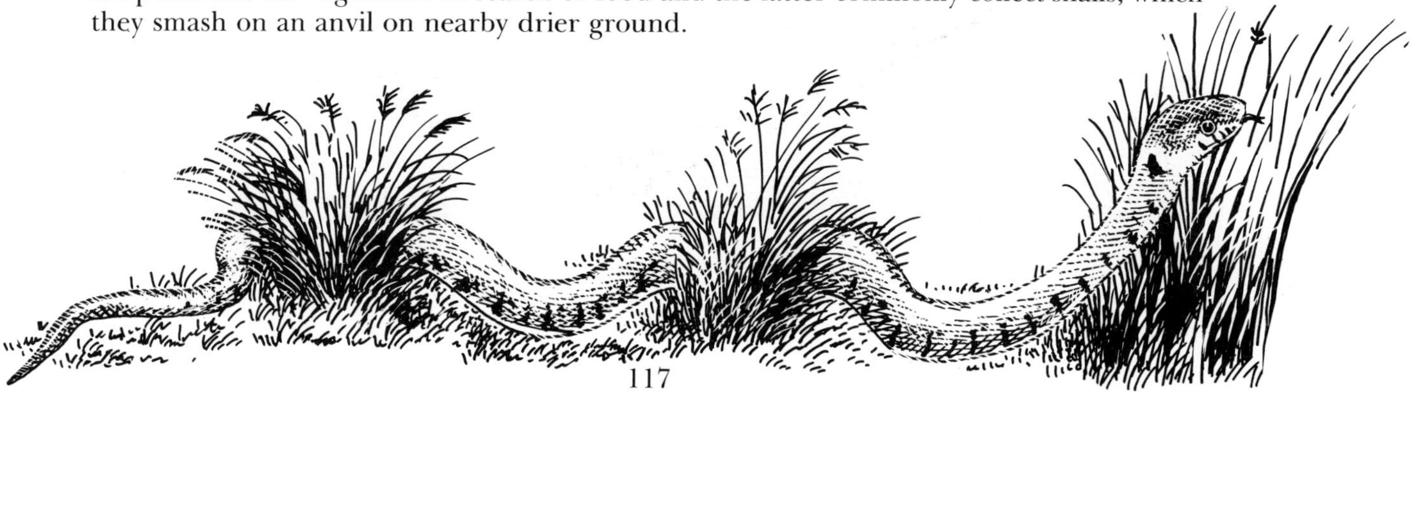

Oak woodland in Wyre Forest. (GHG)

Woodlands

Overall, Worcestershire is not a well-wooded county, although woodland distribution varies from heavily-wooded areas in the north-west Wyre Forest area, to the sparseley wooded south-east, a distribution pattern which has existed for a long time. The WNCT's 1977-78 survey found that 60% (by area) of the county's woods lay in the north-west quarter of the county between Severn and Teme. Elsewhere, Worcestershire's woodland density is well below the national average of 6%, although there are more densely wooded pockets.

Broad-leaved woodlands are the most varied and species-rich of wildlife habitats and, in different forms, are the natural post-glacial climax vegetation. A high proportion of the dry-land fauna and flora is derived from forest habitats, including most of the familiar garden birds, (robins, blackbirds, tits), most mammals, and many hedgerow and even grassland plants and animals are natives of woodlands, woodland edge and clearings.

It is obvious that the original post-glacial wildwood which covered Worcestershire, perhaps as recently as 5,000 years ago, has been cleared or greatly changed. Clearance probably started in Neolithic times, and the pollen record from various periods shows the appearance of agricultural weeds among the tree pollen records. Rackham cites modern studies which show that the most active period of woodland destruction in Lowland Britain was in the late Iron Age and in Roman Britain, over about 700 years. At the end of Roman times about half of Lowland England had probably been cleared of woodland. In Worcestershire, the Vale of Evesham seems to have been cleared at an early stage, probably because the free-draining light soils on the river-terraces were easily cleared and cultivated. The Vale was a corn-producing and exporting area in Roman times and study of the 1086 Domesday survey of Worcestershire shows that it had the highest human population density and the largest number of plough teams in the county. Meadows were also common along the valleys of Avon, southern Severn and the Teme. These indications and the records of woodland

Badger. (BW)

in Domesday indicate that woodlands were widespread in the county, with extensive tracts in the west, north and north-east, in 1086. This confirms the available evidence from earlier Saxon times. At least 25% of Worcestershire was wooded in 1086.

Following the Norman Conquest large tracts of country were added to the Royal Forests and, by the thirteenth century, a greater part, excluding the south-east, fell within one or other of the Forests of Wyre, Ombersley, Malvern, Horwell or Feckenham. Feckenham Forest had been in existence before the Conquest, from 'time immemorial', centred around Hanbury, Bradley, Feckenham, and north to Tardebigge. Post-Conquest it was extended to cover a vast area of north-east Worcestershire, bounded roughly by Worcester, Evesham, Alcester, Alvechurch, Stone near Kidderminster and Droitwich. The Royal Forests were not only woodland: they also contained cultivated land, wetlands, moors, and other open country. The whole Forest was subject to severe forest laws aimed primarily at preserving hunting and timber for the King. The wood within the Forest sheltered wolves, wild boars, foxes, wild cats, martens, badgers, otters, red and fallow deer, all hunted for sport. The 'fowls of the warren' included pheasants, partridges and woodcock. Wolves were abundant in Worcestershire's woods in the Middle Ages. There were small settlements within the the Forests and many people worked in the woods, which provided fuel, pannage and other woodland products. Woodland clearance continued within the Royal Forest with many records of assarting, the term given to clearing areas for agriculture.

During the 1600s the remaining Royal Forests were disafforested. By then there had been severe depredations on both Wyre and Feckenham to provide fuel and charcoal for the demands of the north Worcestershire iron industry, and the production of salt at Droitwich. On disafforestation in 1629, Feckenham Forest consisted of a number of small woods separated by areas of cultivation. However, at about the same time Buchanan quotes one Richard Hill who said that 'Lickhay [Lickey Hills] was full of trees and a squirrel might run a mile and a half in length on the trees without touching the ground, and there were deer on the Lickhay'. Buchanan also extracts information on woodland management from the 1649 survey which gives a glimpse of open woods, for example 282 acre Cornwood, near Newnham, with only nine big trees per acre, and 301 acre Eymore Wood with five. Both woods were worked under a coppice rotation, a commonly used system which for centuries (from about 1200 to 1900, rapidly decreasing 1900 to 1930), supplied large quantities of underwood for fuel, hurdles, wattle and for many other purposes.

Most native broad-leaved trees sprout again after they have been cut down. This commonplace but extraordinary ability, evocatively called 'constant spring' by Oliver Rackham, is the basis of coppice woodland management. Coppicing provides a constant supply of underwood and, after a crop has been harvested by cutting the poles off at ground level, another grows within a few years from the stool. Coppicing prolongs the life of the plant, which develops a bulky, low, multi-branched stump, or an underground root system, called the stool, which can survive for hundreds of years. Woods were either worked as coppice alone or as 'coppice-with-standards' with a low density of trees grown for timber among the underwood. Only a small number of big timber trees can be grown over coppice, as a large number of trees cast shade which depresses coppice growth. A density of timber

Tawny owls live in old woodland. They may be mobbed by small birds. (RMB)

trees of about fourteen per acre of all ages is usual, with only three or four large trees at anyone time. By the beginning of the twentieth century, just before the decline of coppicing, a vast number of specialist uses for coppice products had developed, often with particular uses for particular plants. A common one was to produce 'faggots', bundles of small branches and twigs from coppice or lop and top of big felled trees, for fuel (to produce a quick blaze for heating bread ovens and burning bricks) or for stabilising river banks and coastal defences.

In Worcestershire underwood was used to make the wattle-and-daub panels set between the large timbers of timber-framed buildings. The vertical splints were often oak, with rods of willow, sallow, hazel, ash or occasionally split oak laths, woven between them and the whole covered with a daub of mud with dung or straw. Charcoal was made as an industrial fuel, particularly for iron-smelting, by burning wood in pits or specially constructed heaps, with insufficient air for complete combustion. Wyre Forest supported a large amount of charcoal burning, based on coppiced oak. In the 1890s 'immense forests of oak coppice' in Wyre Forest supplied a thriving charcoal industry. Charcoal made from small wood is fragile and difficult to transport, so iron-smelting furnaces were sometimes built in or near the woods as it was often easier to transport iron-ore to the woods than charcoal to the ore deposits. This is perhaps why Andrew Yarranton's furnace, built about 1650, was set in the depths of woodland near Shrawley. Other underwood products include wattle hurdles, gate hurdles, rake and tool handles, thatching pegs, fence posts, hedge stakes and (hedge top) heatherings. By the 1890s most of Worcestershire's woodland was managed as coppice woodland, with a low density of large timber trees, usually oak, and large quantities of underwood. In the west there was a great demand for ash poles for the abundant hopyards. Ash-beds were common, planted in rows four feet apart, four poles per stool, 8-10,000 poles per acre. Lime and larch also provided hop poles. In the twentieth century the great decline in demand for all underwood products led to great changes in woodland management, and in the wildlife supported by woods. As most of Worcestershire's woods have been used to produce coppice or coppice-with-standards there are now few woods of old trees. This was even so in Lees' time around 1860. In present-day woods, large trees are mainly 80-90 years old, planted at the time of coppice decline.

There has been a growing realisation in recent years that old woods should be conserved and their destruction halted. Their importance lies in the rich, complex and varied communities of plants and animals they contain, which are typical of the natural climax vegetation under the present climate. In Worcestershire it is extremely unlikely that any remnant of 'wildwood', untouched by man, remains, and all woods have been used by people. Nevertheless, many woods are old and grow on sites which have been wooded since trees returned, following the retreat of the last Ice Age about 10,000 years ago. These woods can be traced and listed, the first steps towards their conservation. Modern classifications, like the one used by Tony Whitbread when he prepared an inventory of Worcestershire's ancient woodlands for the Nature Conservancy Council in 1986, place woods in one of several categories. *Recent woods* are on sites which have not been continuously wooded since 1600 AD, and have acquired tree cover, either by planting or natural invasion by trees. *Ancient primary woods* have had woodland cover from at least 1600 AD and have only been used for underwood or timber production. *Ancient secondary woods* have developed woodland on open land at some stage before 1600 AD. In practice the two types of ancient woodland can seldom be definitely distinguished and are generally known as *ancient-semi-natural woodlands.* A further category is that of *ancient replanted woodlands* when a site which can be identified as an ancient woodland site has been replanted with broad-leaved or coniferous trees. Additionally, in central Worcestershire there are a number of woods which have been managed on a unique system, similar to a coppice regime but with considerable inplanting of exotic species among natural broad-leaved regeneration and planting. The aim of this management was to produce large numbers of whitewood poles for turnery. The woods are owned by Harris, the brush makers, of Stoke Prior, although some have been sold and purchased by WNCT and the Woodland Trust in recent years. Although they are in part plantations on ancient woodland sites they are much less damaging to broad-leaved woodland flora and fauna than conifers.

The WNCT 1977-78 survey recorded approximatley 3,000 woods, with a total area of 11,481 ha, 6.6% of the county. Approximately 6,222 ha are thought to be ancient woodland, 3.6% of the county, so approximately 54% of Worcestershire's woodland is on ancient sites. 4,276 ha or 69% of ancient-site woodland is thought to be semi-natural, so 39% of the county's woods can be classified as *ancient semi-natural woodland.* Most of Worcestershire's woods are small, 52% less than 5 ha, 70% less than 10 ha. There are few large woods, with only five greater than 100 ha and only one, the main block of Wyre Forest, larger than 200 ha. Whitbread's inventory of ancient woods, prepared for the NCC, records 6,222 ha of ancient, semi-natural woodland whereas the similar WNCT 1977-78 survey recorded 7,292 ha, the difference being due to inclusion of ancient woodland remnants between 0.1 ha and 2 ha by WNCT.

These figures and much other information were derived from WNCT and NCC field and desk studies, the latter mainly involving the use of First Edition OS one-inch maps prepared in the early nineteenth century and the first series 1:25,000 maps.

The Worcestershire Plain has fewest woods, particularly in the south and east. On the lias clays there tend to be clusters of quite large woods, such as around Grafton, Himbleton and Dormstone and near Wadborough and Drakes Broughton and near the Lenches. These woods are characterised by a mixture of ash, maple and hazel, often with oak which may have been planted in the past, and a rich shrub layer including spindle, dogwood, wild service tree, and midland hawthorn. The ground flora is also quite rich and of the type which grows on lime-rich soils, with bluebells, dog's mercury and herb paris. Few woods exist on the Keuper Marls; these are generally oak and hazel woods such as the Chaddesley and Randan Woods, with sessile oak on the more acidic soils, although small-leaved lime and wild service tree may be present. The most interesting group of woods on the Triassic sandstone is around Shrawley, where there are large woods of small-leaved lime with quite a rich ground flora, including wood fescue, lily-of-the-valley and giant bell-flower. On the Clent Hills, where the soils become more acid, the woods are dominated by oak and often

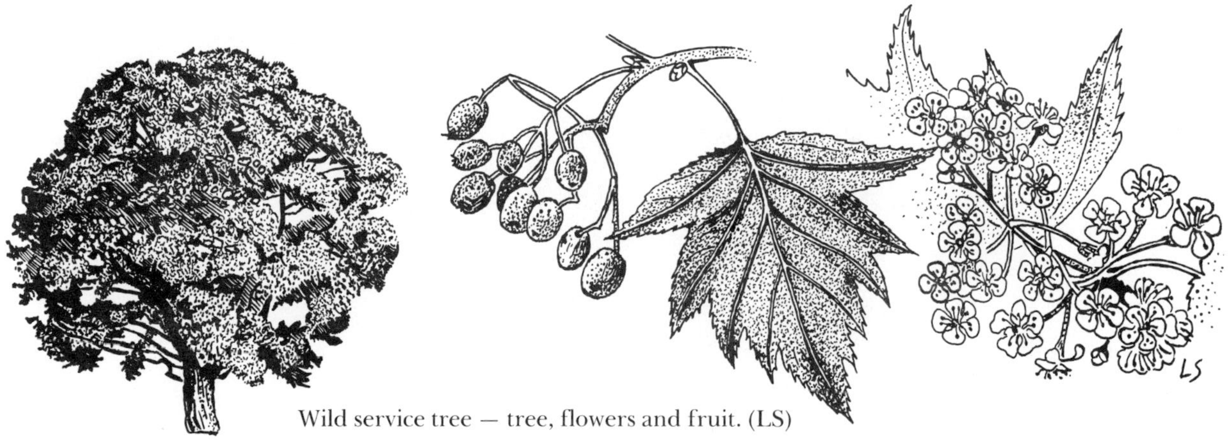

Wild service tree — tree, flowers and fruit. (LS)

contain ground flora species not found elsewhere in the county, including water avens and wood barley. The north-west plateau is dominated by Wyre Forest and this is the most well-wooded area in Worcestershire. The woods are mainly sessile oak over a heathy layer of heather, bracken and greater woodrush, with a poor ground flora, although valley sites along the Dowles Brook are more varied. In the Old Red Sandstone areas of the Teme valley the main woods are on the steep sides of the main valley and in tributary dingles. There is often a gradation of woodland types, from sessile oak and heather on the upper slopes, through wych elm and ash, to alder at the bottom. The West Worcestershire Hills are poorly wooded in the south on the light acidic soils of the Malvern Hills, but more wooded to the north, where there are many species-rich varied woods, often on Silurian limestone.

The loss of ancient semi-natural woods in Worcestershire since 1926 (actually many woods were surveyed by OS long before this date), based on the NCC ancient woodland inventory, has been through clearance or by conversion to plantation forestry. 85 sites have been affected by clearance, covering 394 ha, 76 of them for agriculture. 12 sites, covering 48 ha, have been lost to urban development. 16 sites, 89 ha, have been totally destroyed. A great number of Worcestershire's woods, about 29% and 1,946 ha, have been affected by plantations established on ancient sites. This has been happening since the turn of the century, first with oak, then with conifers, especially in the 1920s after the formation of the Forestry Commission. Most plantations are of conifers such as Scots pine, Corsican pine, sitka spruce, Norway spruce, larch and western hemlock, although broad-leaved species like oak, beech, poplar and sweet chestnut are included in some planting mixtures. The destruction of broad-leaved woods by conifer plantations increased again in the post-war period, and was encouraged in succeeding years by tax incentives rather than good forestry. Many plantations failed to thrive or failed altogether in the face of massive broad-leaved regeneration and lack of forestry management. Properly established conifer plantations, with their deep shade and blanket needle-fall, usually destroy the majority of the fauna and flora of an ancient woodland, although fire-breaks and rides may conserve some natural history interest. The identified loss of ancient semi-natural woodland is about 2,340 ha or 35%.

Until recently, most woodland planting was on sites already wooded, and perhaps felled in normal forestry operations. One striking exception was the establishment of poplar plantations (over 200 ha) on damp valley sites, planted for the match-making industry, before the disappearance of England's Glory and Bryant & May. In their early stages these plantations were often used by marshland birds, such as sedge warblers and, for a few years around 1970, young poplar plantations in the Avon Valley harboured about ten pairs of marsh warblers. When these plantations grow they have little natural history interest, with no ground flora, apart from grass grazed by sheep. Poplar planted on existing woodland sites is less harmful since it allows light to reach the woodland floor and a rich ground flora develops. There are good examples at Eymore and Chaddesley Woods.

LEFT: Sparrowhawk, increasing everywhere. (ARH) RIGHT: Fly agaric, Kingsford. (DH) CENTRE: Bluebells *en masse*, Clent. (DH) BELOW: Land caddis, a rare insect, virtually confined to ancient, semi-natural woodland in the county. (DH)

The other new woods of the county have most often been alder and willow carr, developing naturally on wet soils in river valleys. Some of these woods are quite large. As they mature, the abundance of dead wood and decaying stumps is used by nesting willow tits and lesser-spotted woodpeckers, which excavate their nest holes in rotten branches or thinner trunks. Small willow woods derived from old osier plantations may also provide these facilities, but most have been cleared for agriculture in the last thirty years, especially in the Avon valley.

Woodland surveys have concentrated on woods over 2 ha and there is little numerical information about the loss of small remnant ancient woodland sites. Similarly, there is little information about woods of relatively recent origin, such as the elm spinneys which were common in the countryside before elm disease. Although the south and east of the county have been poorly wooded for a long time, hedgerow elm and spinney elm, abundant until the 1950s, have now vanished. Secondary scrub and woodland arising on rough pastureland have also been cleared.

Wyre Forest is one of the most important wildlife sites in the Midlands. It straddles the Worcestershire-Shropshire border; 921 ha (2,278 acres) are designated SSSI; it is a Nature Conservation Review Site (which means it is listed among Britain's top natural history sites) and part is a National Nature Reserve. It is one of the largest blocks of ancient semi-natural woodland in Britain, comparable in size to the New Forest, the Wye Valley woods and Blean Woods in Kent. Although the area is mainly woodland, there are unimproved meadows, old orchards, small mires, a disused railway line and several small streams. Nearly 1,500 acres are directly conserved by NCC and WNCT through ownership and management agreements. WNCT owns two small reserves, the Betts Reserve and Knowles Coppice, and jointly manages a third with West Midland Bird Club — the Fred Dale Reserve near Dowles Manor. The whole of the greater Wyre Forest block, including associated woodlands, contains nearly 7,000 acres of woodland, but 3,000 acres of this have been converted to conifer plantation.

Wyre Forest lies on a plateau, dissected by streams running into the Severn, and consisting of the varied rocks of the Coal Measures, including sandstones, conglomerates and marls, with occasional pockets of limestone. Soils of the plateau are markedly acidic, while those of the valley sides and bottom tend to be base-rich. The soil variation results in contrasts of vegetation types, often in close proximity to each other. The Forest is a meeting point for many bio-geographical features. The plateau is mainly clothed with sessile oakwood, birch and rowan, and is akin to Welsh valley woods. The oak is derived from old coppice and some large stools appear to be extremely old. There are also areas of pedunculate oakwood like those of the Midlands. The valley woods contain a rich mix of ash, elm, small-leaved lime, wild service and alder, similar to woods of the southern Welsh borders, and the patches of hazel, ash and dogwood suggest East Anglian woods. Other Forest trees include yew, alder buckthorn, holly and wild cherry. The Forest ground flora is also a meeting place for species with differing geographical distributions. For example, the rare westerly mountain sedge grows near the northern mountain mellick. Three species of crane's-bill occur in the Forest: the northern wood crane's-bill, the southern meadow crane's-bill and the bloody crane's-bill, a rare northerly plant of basic soils. The plateau woodland with its acid soils has a ground flora containing heather, bilberry, wood sage, wavy hair-grass, cow-wheat, creeping soft-grass and bracken, and the base-rich, valley woods' soil supports dog's mercury, wood false-brome and woodruff, yellow archangel, wood anenome, violets, pendulous sedge, primrose, toothwort, ramsons and wild columbine. The north-facing slopes of the Dowles Valley are shady and damp, which suits wood-rush, oak fern (the only Worcesteshire site), hard fern and many mosses and liverworts. The south-facing slopes are warmer and drier with saw-wort, betony, bitter vetch, bell heather, and fragrant agrimony.

The old Whitty pear of Wyre Forest. (RMB)

Wyre's boggy flushes are usually base-rich from associated spirorbis limestone and contain such plants as tutsan, hemp agrimony, twayblade, wood horsetail, purple moor grass and occasionally fragrant orchid, great burnet and broad-leaved cotton grass. Acid flushes are rare and may support special plants like bog pimpernel and common cotton grass.

The narrow, or sword-leaved helleborine has long been known as a special plant of the Wyre, where it grows, somewhat uncharacteristically, beneath oak coppice.

The tale of the sorb tree, or whitty pear, *Sorbus domestica,* has to be included in any account of the Wyre Forest. The solitary tree in the depths of Wyre Forest was first mentioned by Alderman Edmund Pitts who wrote about it in 1678 as 'a rarity growing in the forest in this county of Worcester'. It was an old tree even then but it survived until 1862, when it was burned down. In the early 1820s, two cuttings from the tree had been rooted at Lord Mountnorris's arboretum at Arley Castle, where they still remain in the grounds of a private house built on the site of the old castle. A sapling raised from one of these trees was eventually planted on the site of the old tree in 1916, and it is still there with a small memorial plaque. The history of these trees is closely bound up with the history of the Worcestershire Naturalists' Club and a full and delightful account is written by Mary Jones in her book *The Lookers-out of Worcestershire.* The tree has leaves rather like a mountain ash and bears fruit similar to those of the wild service tree, *Sorbus torminalis,* except they are pear-shaped rather than round. Rowans were called 'whitty trees' and associated with witches, as they were supposed to give protection against bewitchment. The solitary 'whitty pear' in Wyre, with rowan-like leaves and small, bitter, pear-like fruits instead of the rowan's scarlet berries, was apparently even more efficacious. Several sorb trees are now growing elsewhere in Worcestershire, thought to be offspring of the Arley trees. The species is a native of southern and eastern Europe and the origins of the Wyre tree are unknown, although Edwin Lees, writing in 1967, speculates that it was planted or grown from seed on the site of a hermitage which perhaps had a garden, as plum and privet were growing nearby.

Wyre Forest, with its mixture of rides and clearings, many varieties of woodland and abundant flora, is excellent for insects, especially butterflies and moths. Of the 45 species of butterfly recorded in Worcestershire in the last twenty-five years, thirty have been seen in Wyre. These include five species of fritillary whose caterpillars are dependent on a good growth of violets, which thrive in coppice rotations. The masses of bramble blossom, clumps of thistle, knapweed and hemp agrimony flowers are attractive to many species of butterfly. The big, fast-flying fritillaries are often joined at the flowers by commoner red admirals, peacocks, skippers and browns but the small purple hairstreaks, with a flash of purple on the forewing and grey underwings, are elusive because they spend most of their time among the topmost twigs and leaves of oak trees. They over-winter as eggs and the young woodlouse-shaped caterpillars hatch in time to eat oak buds and then leaves. Pupae are placed on the tree trunks and butterflies hatch to be on the wing July to September. But you need binoculars and sharp eyes to see them in the tree tops.

Moths are abundant and hundreds of species have been found, many of them local and rare species. The Forest is important as the only English site for the Kentish glory, although

Great-spotted woodpeckers extract wood-boring insects from dead wood. (LS)

it has not been seen for some time. The caterpillar feeds on low birch scrub and it is possible that encouragement of such habitat by new coppicing would benefit the moth. There are also many northern and western species like the alder kitten, angle-striped sallow moth, and the sword-grass and red sword-grass moths which are principally found in northern Scotland. Heathland moths associated with heather and gorse, like true-lover's knot and grass wave, are also found.

Another group of striking insects are the wood-wasps which are often large, black and yellow. The female has a stout ovipositor (egg depositor) which is used for drilling holes in trees during egg-laying. The larvae bore their way through wood as they grow. The larvae of one large species, *Uroceras gigas* are parasitised by an ichneumon (parasitic agent) with an extremely long and fine ovipositor. *Rhyssa persuasoria* detects the larvae deep within the wood by careful search of a suitable tree and then inserts the ovipositor through the timber. The egg is laid on the larvae and when it hatches the Ichneumon larvae consumes the host. Egg-laying is a remarkable procedure to watch — *Rhyssa* almost stands on its head when it starts to work the ovipositor into the wood, guided by the back legs. The fine tube, about 3 cms long, is then gradually drilled into solid wood over a period of minutes.

The large nests of wood ants are common in the Forest. These are heaps of fragments of vegetation and in warm weather swarm with large workers over 1 cm long.

Wyre is one of the few British localities for the terrestrial caddis fly, rediscovered there in 1957 by Dr Norman Hickin and his young daughter, one hundred years after the first record. This is a most unusual species; the larvae live among damp woodland floor vegetation and eat dead oak leaves, quite unlike all other caddis fly larvae, which live in freshwater. Like many of the latter, the terrestrial caddis larvae builds a protective tube of small plant fragments in which the body is encased. Adult male caddis flies have hairy soft wings and frequently fly over water.

A great many species of birds are found in Wyre, including the typical 'western wood' summer visitor species — redstart, pied flycatcher and wood warbler. Wood warblers are found in areas with taller oaks which are fairly open beneath the tree canopy. There are more wood warblers in Wyre than anywhere else in Worcestershire and the dual songs (one a rapid trill, the other slower and lilting) are a feature of May days. The birds spend most of their time in the canopy, although they nest in low cover on the ground. Both pied flycatchers and redstarts are hole-nesting species, competing with the tits. The former is often associated with old orchards which are dotted along the Dowles Valley. Great-spotted woodpeckers are common throughout the broad-leaved woodlands, lesser spotted are often found along the stream valleys, where there are alder and small branches of rotten wood. Green woodpeckers bore nest-holes in living timber and often feed on ants among the anthills which are abundant in the little meadows round Wyre. All the woodpeckers are much dependent, especially in winter, on a good supply of dead wood containing wood-boring insects and their larvae. The woodpeckers' nest-holes are also used by starlings and

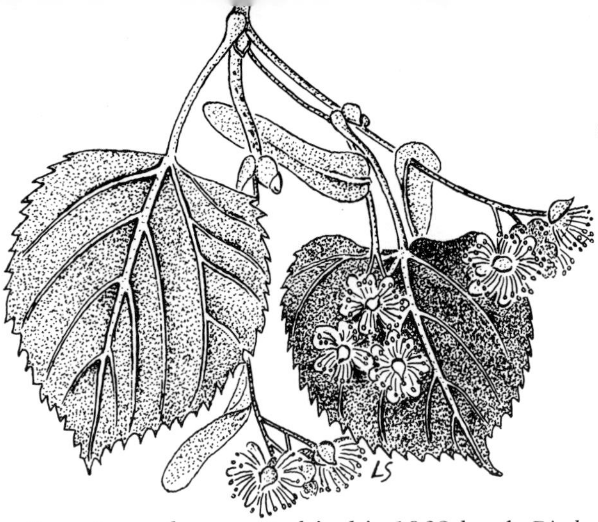

Small-leaved lime. (LS)

nuthatches. Woodcock are a feature of Wyre, although scarce in Worcestershire generally, and can be seen in territorial roding flights over the Forest in spring and summer. Hawfinches breed, tawny owls are thickly distributed and the woodlands are hunted by sparrowhawks. Wyre has another claim to ornithological fame: Brunsdon Yapp carried out some of the first research into the estimation of bird numbers in woods using transect methods in Wyre Forest, partly reported in his 1962 book *Birds and Woods*.

Most common mammals occur in Wyre, and some of the more uncommon species. Red squirrels used to be common but have now disappeared and been replaced by the ubiquitous grey squirrel, introduced into England from North America in the nineteenth century. Hedgehogs, moles, wood mice and bank voles are common. Foxes and badgers are frequent and otters were once common on the Dowles and appear to be returning. Dormice live in Wyre and appear to be quite common, although the species is declining in many parts of England. They are stout, furry and with a long, rather bushy tail and are largely nocturnal. The nest is usually in a hole and often constructed of stripped honeysuckle bark. Dormice (and other mice) quite often take over nestboxes put up for birds. Their food is nuts, berries and seeds. Fallow deer are present in the Forest and are a problem to foresters and those who want to re-introduce coppice systems. Deer are grazers of young woody shoots and prevent regrowth of trees and understorey. The Forestry Commission endeavour to control numbers by shooting: a practice which is necessary for the future good of the woodland as large numbers of deer do a great deal of damage. At WNCT's Knowles Coppice reserve it it is hoped that deer fences will allow regrowth of oak and other trees from seedlings, which are abundant. Glades are being felled for this purpose, as it is most important to produce a stand of trees of different ages, to ensure the continuing supply of oaks in the future.

Native lime trees (not the hybrid commonly planted in towns and parklands) occur in Worcestershire in considerable numbers. The small-leaved lime is found in many woods in west Worcestershire with two other distinct concentrations in Lincolnshire and southern East Anglia, and smaller ones elsewhere. Large-leaved lime is much more uncommon in England but does occur in small numbers in Worcestershire's woods, mostly in the west. According to the pollen record, small-leaved lime was abundant about 5,000 years ago with a similar, but less discontinuous, national distribution to the present day. A lime decline occurred in Neolithic and Bronze Age times, probably due to increasing agriculture. Lime may have been fairly easy to kill as it is palatable to grazing animals: cattle will eat both leaves and bark. One particular use of lime was the preparation of a poor-quality string from the bast — a fibrous layer on the inner side of the bark. Lime quickly produces poles when coppiced. The timber is soft, relatively grain-free, and ideal for wood carving.

Shrawley Wood, on the west bank of the Severn about three miles south of Stourport, is one of the largest and most important limewoods in England. Many neighbouring woods also contain lime and Lees, writing in 1862, states that lime appeared in numerous woods, coppices, and old hedgerows from Wyre Forest south to the end of the Malvern Hills at Bromsberrow. He also mentioned Round Hills near Abberley, Ockeridge Wood, Berrow Hill near Martley, the banks of the Leigh Brook, Rosebury Rock (which seems to be called 'Osebury' Rock nowadays), along the Teme, Old Storridge Hill, about Great Malvern, and in ancient woods at Castlemorton and Berrow. Certainly some present-day woods in this area contain small-leaved lime, but it now seems less abundant. Lees also mentions that many old lime trees were pollarded and in time became grotesque. A tree meeting this description was recently noticed in Hallhouse Wood, near Suckley.

In Lees' time Shrawley Wood extended over 500 acres of lime copice. It is now less extensive, having been partly cleared and partly planted with conifers. Nevertheless, 255 acres are now an SSSI and the wood is one of Worcestershire's Nature Conservation Review sites and of national importance. It is partly owned by the Forestry Commission and part is in private hands. The latter section has been continuously managed as lime coppice for centuries. When the FC purchased a section of the wood, conifers were planted on parts, but fortunately large areas are still covered with lime, partly as coppice and partly as 'singled high forest', where single stems from the old coppice stools have been grown on as large trees. The site lies over Keuper sandsones, a little above the terraces and flood plain of the Severn. It is freely draining and the soils are weakly acidic with base-rich flushes in the valleys. There are great carpets of bluebells beneath the lime in many areas and bracken in others. Foxgloves and wood sage grow on the acidic soils and dog's mercury, enchanter's nightshade and ramsons are abundant in the basic areas. The many interesting and unusual plants include giant bellflower, spreading bellflower, broad-leaved helleborine, lily-of-the-valley and herb paris. Worryingly, Himalayan balsam is creeping up the streams from the river Severn and into parts of the wood.

The older coppiced areas contain a few standard trees, mainly large oak, small-leaved lime and sweet chestnut. Huge oaks from Shrawley Wood were used to re-panel the House of Commons after World War II. These great trees tower over the coppice. After cutting, the lime-coppice stool produces a mass of stems which grow into a bushy structure ten feet high within about four years. During this period the ground between the bushes is filled with bramble, newly germinated birch scrub and rosebay willow herb. This habitat is rich in birds with large numbers of breeding warblers. As the coppice grows, the stems 'self-thin' and reduce in number and by the eighth year most of the canopy closes and dense shade suppresses the undergrowth. Some birches grow as quickly as the lime and keep pace with it for years. Evidence of similar behaviour by birch in the past is seen by some massive birch trunks eighteen inches in diameter in forty-year-old coppice, which had presumably been left to grow when the coppice was last felled. Many of these died in the great drought of 1976. Forty-year-old lime coppice forms an extraordinary wood with a relatively simple ground cover of spring-flowering bluebells and ferns. The lime poles, perhaps a dozen from each stool, are fifty feet tall and, when in leaf, cast heavy shade. The singled coppice of trees about seventy years old and seventy-eighty feet tall casts less shade than old coppice and the ground flora is often dominated by a low tangle of brambles. The flowers of the lime canopy are rich in nectar and in summer hum with a myriad of bees and hoverflies.

A detailed study of the birds of these types of lime wood has been running for seven years. When it started in 1984 the tall, unthinned forty-year-old coppice was found to be fairly poor in birds, with an overall density of 36.5 pairs/10 ha. The following year it had fallen to only 18.4 pairs/10 ha due to a decrease in robins, wrens and blackbirds following cold winter weather. Since then the population has increased every year to 53.1 pairs/10 ha in 1990, the increase being almost entirely of the resident, non-migratory robins, wrens, blackbirds and chaffinches. This increase follows a series of mild winters without hard frost or long snow cover, which kills small birds by cold and starvation.

The younger coppice, five years old when the study started, contained 76.5 pair/10 ha at first, over twice the density of the old coppice. There were many summer immigrant warblers, garden warblers being most abundant, followed by willow warblers, chiffchaffs, blackcaps

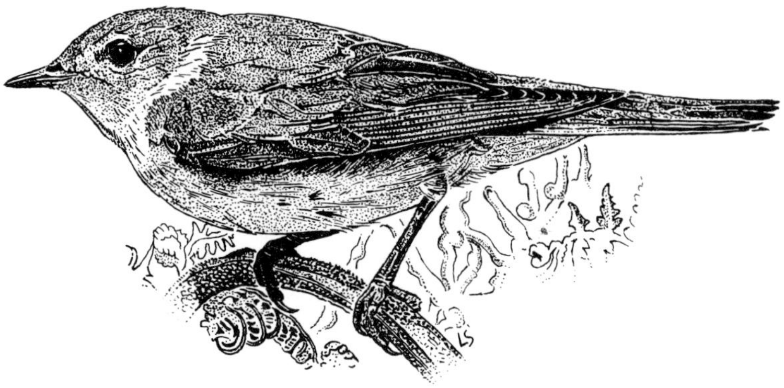

LEFT: Wren. RIGHT: Garden warbler — a bird of growing coppice. (WNCT)

and whitethroats, together with common resident and scrub birds yellowhammer, bullfinch and long-tailed tit. As in the old coppice there was a decrease in wrens, robins and blackbirds the following year, following the cold winter. In the third year numbers reached 83.4 pairs/10 ha, the highest figure recorded. Residents had increased a little, as had willow warblers, and garden warblers had declined. In the subsequent years to 1990, garden warblers have decreased from 12.5 pairs/10 ha to 7 pairs/10 ha. Blackcaps have also declined. Yellowhammers, birds of low scrub, have disappeared and turtle doves, summer visitors using tall scrub, have increased. Resident birds have all increased (the mild winters) and all the changes have balanced out to an overall bird population of 76.5 pairs/10 ha, as at the start of the study. As the coppice becomes older and shades out the ground cover, the summer warblers decline, and observations in a coppice a few years older within Shrawley Wood suggests that warbler numbers will drop sharply in the next few years. In both of these coppice areas (young and old) the numbers of tits recorded depends on the presence of the huge, old standard trees which tower above the coppice and provide nest-holes and abundant insect food for the nestlings. These trees contain large holes with nesting jackdaws, stock doves and kestrels.

The singled-lime woodland falls between the two coppice plots in breeding bird numbers. The overall density at the start of the study was 38.6 pairs/10 ha which fell to 29.3 pairs/10 ha the next year, after the cold winter which depleted wrens, robins and blackbirds. Thereafter the population has increased steadily to 56.8 pairs/10 ha due to increases in robins, wrens, blackbirds and chaffinches. Chiffchaffs have moved into the singled woodland. This species is almost always associated with big, mature trees (as it is in the other plots) and perhaps the maturing lime wood is more to its liking.

The changes in bird population in relation to coppicing highlight the importance of the system to wildlife. The young coppice provides breeding habitat for many resident and summer visitors and becomes less suitable as the coppice grows. When it is felled the cycle starts again. If a wood is worked annually with coppice plots in different parts of the wood, the system will provide a continuing variety of habitats for a great number of species.

Invertebrates are greatly influenced by coppice cycles. For example, fritillary butterflies rely on violets as a food plant and the virtual disappearance of coppice in Worcestershire and elsewhere is probably the main reason for the great decline in these butterflies. On the other hand, white admiral butterflies have actually increased in Worcestershire, and spread north, because of the decline in coppicing. This has encouraged greater growth of the caterpillar's food plant, honeysuckle, in woodland shade. The eggs are laid on dangling wisps of honeysuckle in dappled shady areas. In August the caterpillar eats the sides of the leaf and leaves the midrib sticking out. It then hibernates, emerges the next year to eat

RIGHT: Treecreepers search for small insect prey in the crevices of bark. (RMB) BELOW: Blackcap — bird of mature coppice and thickets. (WNCT)

honeysuckle leaves until May, pupates for June and the butterfly hatches in July. During the first year or two after coppicing the ground is exposed to sunlight and becomes warm, good habitat for many wolf and jumping spiders which hunt over the ground for other insects. Wood ants are also common then and these-sun loving invertebrates move elsewhere as coppice grows and shade increases.

Plants respond dramatically to coppicing. In the first few years there may be a great show of spring flowers, such as primroses, wood anemones and yellow archangel, which survive under dark, older coppice without flowering. Other plants grow from the seed bank in the soil, or from seed which reaches the area when coppice is cut, and can be very varied, with rushes, St John's-worts, foxgloves or wood spurge.

Small mammals are also affected by coppice, with great increases in wood mice in the second or third year. Dormice, if they occur, need a closed coppice canopy in which to climb, and preferably a supply of nuts from hazel which fruits after about six years' growth.

Many of the woods on the Worcestershire plain were once worked as hazel or mixed coppice, with or without oak standard trees. Most are now little used for forestry, but often for pheasant shooting. In many cases the old coppice woods were planted with more oaks. These are now around 70 years old and the tree canopy has just about closed, casting considerable shade and suppressing the coppice beneath.

Hornhill Wood, a small 14 acre wood near Himbleton, is in some ways an animated fossil! Here a classic rotation system of coppice-with-standards has been re-instituted by WNCT. The standard trees are mainly pedunculate oak with an understorey mainly of hazel. Records state that a ten-year coppice cycle was in use in the wood in 1649, still practised in 1777 and continued less effectively into the 1960s. WNCT has re-cycled the wood once and is now on the second rotation. The aim has been to raise the number of stools to the classic 1,200 per acre and this has been achieved by planting and layering. The original coppice cycles required sufficient new timber trees to be selected each time the coppice was felled,

with a target of about 34 young oaks per acre, through all age classes to three big 200-year-old trees per acre. This had not happened at Hornhill and twelve years ago most of the trees were 60-80 years old with almost complete canopy closure, shading the coppice and suppressing growth. WNCT has been felling some trees to open up the canopy while planting new trees of several species to create a wide age range for the future. The rewards for these efforts have been considerable, when compared with nearby woods of similar tree age and structure. In the second year after coppicing there is a magnificent display of primroses and violets and for the next year or two the number of flowering plants

ABOVE: White admiral has spread into the county in derelict coppice woods rich in honeysuckle. (WNCT) BELOW: Silver-washed fritillary. (LS)

increases steeply, before declining again as the developing hazel canopy casts a deep shade. In the last few years of growth numbers of garden warblers and blackcaps increase, although the small plots have only small effects on the wood's warbler populations. The work in Hornhill has been done almost entirely by volunteers: the oaks have provided many long-lasting cleft fence posts for reserves, and the hazel an abundance of pea and bean sticks and pegs for thatching.

Trench Wood, near Huddington, and Monkwood, near Grimley, are two large woods originally managed by the 'Harris Brush regime' mentioned earlier. Trench Wood was once rather like Hornhill Wood, a standard oak wood over hazel coppice although much richer in wildlife since it is situated on clays and base-rich soils, unlike the rather poor Keuper Marls underlying Hornhill Wood. 'Trench Wood' is three contiguous woods, two called 'The Trenches' and Popeshill Wood. The parish boundary runs between the Trenches and along the outside of Popeshill Wood and is clearly marked by wood-bank and ditch. The north-west part of the Trenches is in private ownership and consists of tall, roughly 100-year-old oaks, with a closed canopy and rather straggly understorey of hazel and other shrubs. Like many woods in the district it is a 'derelict' coppice-with-standards wood which is naturally developing into high forest. There are many dead branches and tree-holes and this part of the wood is excellent for hole-nesting tits, woodpeckers and tree creepers. It is also part of a pheasant shooot. There is a similar small block of woodland in the south-east. The remaining 100 acres of the wood was 'Harris-ed' for about twenty-five years before purchase by WNCT. This entailed clearance of all big timber trees, allowing natural regeneration, selection and planting in trees to provide a quick growth of white timber for turnery which led to a young woodland of many species growing closely together – birch, Italian alder, beech, and sycamore.

Trench Wood has been known as an interesting wood with many rare plants and insects since at least 1834. The great stirring-up caused by the Harris activities brought forth an excellent showing of ride plants with great butterfly and common spotted orchids among them. Many butterflies have been recorded and this wood is the last site for marsh fritillary in Worcestershire, the caterpillars feeding on devil's-bit scabious in the rides. The centre of Trench Wood has been poorly wooded and now supports thick, low, mixed scrub. These scrubland areas and the Harris plantations are good for breeding birds, especially warblers and nightingales. Worcestershire lies on the north-western edge of the nightingale's range in England and the numbers breeding in the county vary from year to year. The birds may use a wood one year and desert it the next. Trench Wood has supported consistently good numbers for many years with eight to twelve singing birds. The structure of the young wood, with many small trees of twelve to fifteen feet high, moderate light penetration and a tallish understorey with bramble and rose, seems to suit these birds. However, they are all eclipsed by willow warblers whose territories completely fill the wood. In 1987 there were at least 63 willow warbler territories in a 33.5 ha census plot, a density of 18.8/10 ha. The non-migratory wrens, robins, and blackbirds and summer immigrant garden warblers and blackcaps were all

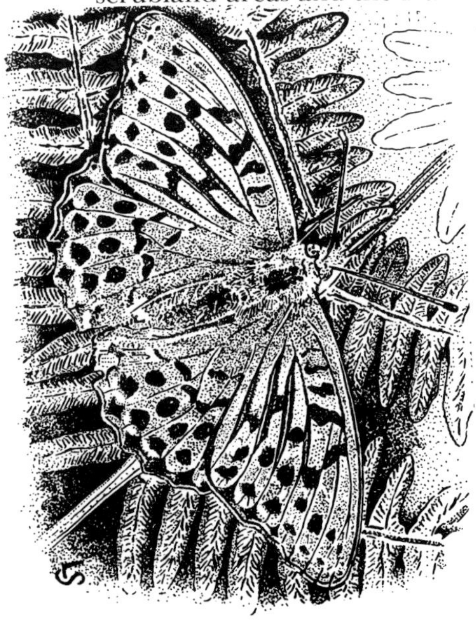

ABOVE: The rare wood white is found in Monkwood. BELOW: Purple hairstreaks live in the tops of oak trees in old woods. (WNCT)

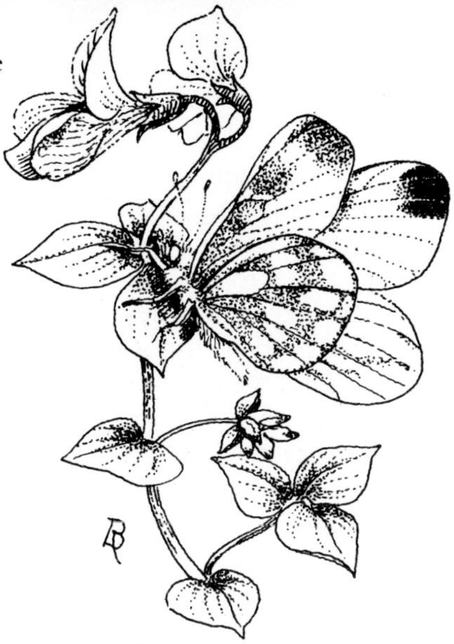

present in densities similar to those in the Shrawley Wood young lime coppice. Interestingly, there were relatively few chaffinches and chiffchaffs in the 'Harris-ed' plot, as could be predicted because both species are associated with trees during the breeding season. Few hole-nesting tits were present, but great-spotted woodpeckers nested in the birches. In winter large flocks of redpolls and siskins may gather to eat seed from the alder false-cones.

Tiddesley Wood, on the outskirts of Pershore, an ancient woodland site, was purchased by WNCT on 1984. It was recorded in Domesday Book in 1086 as 'woodland 2 leagues long and 3 furlongs wide' near Pershore. Research by Dr Pam Gough has revealed the first written records of a 'Tiddesle Wood' in 1218 and the Saxon suffix 'lea' may describe a wood surrounded by cleared land. The wood was commonland and grazed by the commoners' cows and pigs in the autumn until about 1223. The Abbot of Westminster then induced landowners to give up their common rights so he might empark it. In 1271 William, Earl of Warwick, gave the Abbot of Westminster permission to *reinclose* his park at Tiddesley perhaps 'after hedges and ditches had been thrown down and his property damaged and his animals carried off' apparently by tenants of the Abbot of Pershore who continued to claim common rights. The purpose of the park was to breed deer for food.

Dr Marshall Wilson has recently found documents in the Westminster Abbey archives which increase understanding of the way in which Tiddesley Wood was managed in the past, and perhaps provide guidelines for future management as a nature reserve. A rather crude map of 1708 shows the shape of the wood, including the meadows on the west side alongside Bow Brook and a field cut into the east side. A fine map by Joseph Dougharty of a family of well-known Worcestershire map-makers, dated 1736, shows a wood shaped similarly to the one shown on first editon OS maps.

A survey document of 1813 states that the wood consisted of various sorts of underwood, such as oak, in considerable quantity, with some ash, maple, sallow, hazel etc. It was well stocked with young oak, termed black poles and usually felled at 39 years, and white poles felled at 26 years, and coppice at 13 years. The black and the white poles were valuable for their size and bark. There is also a letter of explanation, dated 1824, quoting a report of 1806 which enlarges on the coppicing arrangements: the wood was divided into falls and, when the ordinary coppice was cut at 14-16 years, 150-200 oak poles of that age were marked, called tellers, and left to grow for another 14-16 years. Half of these were then cut (14-16 years later) with the small coppice, being 28-32-years-old and called white poles. 14-16-years later still, when the small coppice was cut again, the rest of the poles were cut at about 44-48-years-old, and called black poles. At each cutting care was taken to mark more poles to ensure the continuing succession. Because the oaks were all grown from stools they were not classed as timber, otherwise they would have been the property of the landlord. Another document, a lease dated 1862, prevents cutting of underwood of less than twelve years' growth.

These documents show that Tiddesley Wood was worked on an unusual type of coppice cycle, which allows quite high density of oaks (50-75) to reach 42 years, and none to a greater age. At the same time the successor oaks were growing on *and* the shrubbery beneath was harvested every 14 years.

Unfortunately we do not know how many 'falls' there were in the wood and how many were cut each year. Nevertheless Tiddesley must have produced a great deal of underwood for faggots, wattle, and similar purposes and quite well-grown oak for constructional work, maybe charcoal, and to provide bark for tanning.

In 1952 Tiddesley Wood was leased to the Forestry Commission who, in the great post-war drive to grow timber quickly, attempted to ignore history and convert Tiddesley Wood into a conifer plantation. They waged chemical and physical warfare on the wood in an attempt to kill the broad-leaved trees and shrubs, and then planted Scots pine, Corsican pine and Douglas firs, with a small number of other species. Oak and beech were on the planting plan and a few beech, not native to the site, survive. The grand coniferisation plan largely failed, presumably because resources were short and labour unavailable to cut back all the rapid broad-leaved coppice growth, which overwhelmed the conifers in many parts of the wood. The planting programme was not completed but, where the conifers were weeded and cleaned by the foresters, they grew well and eventually their shade and needle-fall suppressed the broad-leaved hardwoods. By the 1980s WNCT had discovered that Tiddesley Wood still contained a rich ground flora and was an ancient woodland site worthy of conservation. The government of the day required the FC to dispose of many woodlands and WNCT was fortunate to receive many grants and donations and to be able to purchase it. All but a small part (14 acres are in private ownership) is managed as a reserve and the whole wood is scheduled as an SSSI.

Tiddesley Wood has a splendid display of spring celandines, bluebells, primroses, cowslips and wood anemones, followed by bugle, wood spurge and many others. Herb paris is abundant, mixed with adder's-tongue fern, bird's-nest orchids, great butterfly orchid and spotted orchid. The lime-rich soil supports swathes of dog's mercury, and patches of twayblades. Violet helleborines lurk in dark areas and the true autumn crocus *Colchicum autumnale* flowers in September. Scented agrimony lines parts of some paths and great clumps of pendulous sedge are found in wet areas. There are even dry patches with bracken and large patches of ivy. The large amount of ivy on the woodland floor in places is puzzling in an ancient wood. It is possible that ivy seeds were introduced in large quantities when it was occupied by a huge starling roost twenty years ago, or by wood pigeons which also once roosted in the wood in large numbers.

Following the FC's clearance there was massive regeneration of many coppice and scrub species to form thickets with an enormous range of trees and shrubs: oak, ash, wych elm, smaller leaved 'lineage' elm, maple, hazel, small-leaved lime, alder, aspen, birch, sallow, various willows, crab apples, wild pear, dogwood, elder, blackthorn, woodland hawthorn, common hawthorns, stands of wild service trees, brambles, dog rose, wayfaring tree and various wild plums. The original famous yellow egg plum of the Vale of Evesham was found as a seedling in Tiddesley Wood in 1833. Wild clematis and honeysuckle scramble over the trees and many a poor specimen of a conifer has formed a spiral twist to the trunk caused by strangling honeysuckle. In recent years the dense thickets have started to clear through the effects of shading and competition, leaving a more open, young woodland. Parts of this are now being converted to cycles of coppice and coppice-with-standards plots and other areas are being thinned to clear the canopy to give space for the best trees of many species to grow.

The system of wet and dry, and broad and narrow paths and rides contains many grassland and marsh plants (ragged robin, great hairy willow herb, meadow sweet), and these and the flowers of bramble, marsh thistles and teazels attract flies, bees, bumble bees, hoverflies, wasps hunting for prey and unusual beetles and butterflies. The wood has an excellent and varied population of butterflies. White admirals have increased, presumably because of the ready availability of dangling honeysuckle in shady places for the caterpillars to feed upon. Purple hairstreaks are seen high on the oaks. There is a good colony of white-letter hairstreaks and the caterpillar can be found eating elm flowers and leaves, although the butterflies themselves are elusive, spending much time perched high in elm trees, but occasionally descending to flowers. Brimstones often feed on the thistle and teazel flowers where they are often joined by peacocks and small tortoiseshells. Small coppers enjoy the late flowers of fleabane in the wetter rides. Late commas dangle from the flowers of devil's-bit scabious and even later red admirals visit ivy flowers or suck juices from the fallen wild pears. The wood is visited by many white-legged damzel flies and southern hawker and brown hawker dragonflies and banded agrions from the nearby River Avon and Bow Brook.

Tiddesley Wood suffers from two pest species — rabbits and grey squirrels. As everyone knows, rabbits were a scourge of agriculture, causing great crop losses before the appearance of myxomatosis which killed them by the million. Nowadays local populations do build up and by 1990 there were great numbers of rabbits in Tiddesley Wood and grazing crops on adjacent fields. In this situation WNCT is required by law to take steps to stop the damage by reducing their numbers or by fencing. Within the wood rabbits create short grassland habitats and eat almost anything. They seem to be especially fond of the flower buds of twayblades. Former warrens are covered by elder scrub with ground ivy beneath as these are both plants which rabbits do not eat. The other pest is the grey squirrel. Unfortunately these attractive animals attack the growing points of many species of tree, distorting them and preventing the growth of good timber. They also eat a great many young hazel nuts; crops are usually completely destroyed before they ripen. The squirrels build a large ball for a nest, the dray, high up against the trunk of a tree, in which the young are raised. Fortunately, there are as yet no deer in Tiddesley Wood for, although they too are appealing animals, it is almost impossible to work coppice regimes when they are abundant, because they eat the new shoots.

Few birds breed in Tiddesley Wood. Small numbers of garden warblers and blackcaps nest as do nightingales in some years with a few pairs of chiffchaffs. Willow warblers are absent; presumably the woodland structure is not to their liking, although the new coppice regimes may attract them. Breeding tits are scarce, reflecting the lack of old, holey timber for nest sites but nest-boxes are rapidly occupied. Marsh and willow tits breed in a few places, the latter excavating a hole in a rotten tree stump. Sparrowhawks hunt the wood by day, and a pair of tawny owls by night. The conifer plantations add another dimension to the wood's birds as they are attractive to both coal tits and goldcrests. The latter are scarce in broad-leaved woods although coal tits are common. Goldcrests build their tiny nests slung from the thin branches of conifers and obtain most of their insect food from twigs and needles. Coal tits often nest in holes in the ground, or near the bases of trees or in an old stool. In the years when crossbill populations erupt from northern Europe they are usually found in the wood in autumn and winter, extracting seeds from spruce cones. Other winter birds include foraging parties of tits moving through the trees and parties of siskins and redpolls taking seed from the alders.

Since 1985 WNCT has started new coppice systems in parts of Tiddesley Wood and devised a programme of thinning to develop a broad-leaved high forest woodland in others. The conifers are being removed, except in the areas in which they have thrived. These areas are being thinned for eventual cropping. Rides have been opened and widened. In time the diversity of habitats maintained by all these operations will conserve the richness of the wood and show that woodland conservation, economic forestry and visitors are all compatible.

Lion and Brockhill oakwoods lie on the Keuper Marls at the county boundary with Warwickshire near Wythall. The soils of both woods are derived from glacial drift leached soils. In Lion Wood these conditions have encouraged the development of a calcifuge ground flora containing wood sage, wavy hair-grass, heath bedstraw, wood sorrel, foxgloves, and bilberry, with extensive areas of bramble. There is an interesting shrub layer of rowan and holly, with many alder buckthorn bushes and small trees, all providing an excellent crop of berries for birds. The trees are mainly big oaks interspersed with occasional large rowans and hollies and patches of birch. Lion Wood is bordered by a small field which contains calcifuge plants such as lesser spearwort and sneezewort. Brockhill Wood, only a few yards away, used to be similar to Lion Wood. In 1981 the underwood was cleared, the ground flora bull-dozed, and many big trees cut down, thus converting it to wood pasture. Lion Wood is a nature reseve managed jointly by the Warwickshire Nature Conservation Trust and WNCT.

The West Worcestershire woods fall into two main categories: the dingle woods of the Teme valley and the woods clothing the hills north and south of the Knightsford Gap. South of the gap there are extensive woods near Lulsley (the Ravenshill Woods), on the Suckley Hills, at Old Storridge Common, in the Leigh Brook valley, at Halesend Wood and between Storridge and West Malvern. Most of these woods are on base-rich soils derived from Woolhope and Wenlock limestones of the Silurian rocks. Many are on ancient woodland sites and have at one time been managed by coppice-with-standards regimes. They contain a typically rich mix of tree and shrub species including oak, small-leaved lime, birch, field maple, wild service tree, wild cherry, wych elm, elm and yew, and the ground flora associated with limestone soils: dog's mercury, bluebells, wood anemone with nettle-leaved bellflower, toothwort (a semi-parasite often on hazel roots), wild daffodils, early-purple orchid, broad-leaved helleborine and sanicle. Hallhouse Coppice and Suckley Wood have had a more chequered history as they were extensively replanted with conifers in 1964 and 1970 with mixed success. Crews Hill Wood SSSI at the northern end of Suckley Woods gives a glimpse of former glories. Here the main canopy tree is sessile oak, associated with hornbeam and ash in the northern part. In other parts, the oak and ash are associated with three other distinct communities: one, beech, birch, large-leaved lime and wild service; two, small-leaved lime and wild service, and three, wild service and yew. The presence of an oak-hornbeam wood in Worcestershire is unusual. At least twenty-three species of tree, shrub and climber occur in the wood, together with a typical limestone ground flora. Apparently much of Suckley and Hallhouse Woods were similar in nature before they were cleared for attempted coniferisation. Suckley Wood is dotted with large, dead yews which, at the time of clearance, were ring-barked or poisoned with diesel oil. The conifers have failed over much of Suckley Wood and there is a mixed regeneration of young broad-leaved trees and interesting ride-side flora.

Similar types of woods occur north of the Knightsford Gap towards Abberley Hills, with some interesting ones tucked away in small valleys and on ridges. The larger Elbatch Wood received the Harris-treatment, as described for Trench Wood. There are extensive conifer plantations on Woodbury Hill and on parts of the Abberley Hills. The latter bear a strip of ancient woodland on the north side, including extensive stands of yews, with an abundant ground layer of dog's mercury. Early purple orchids and narrow-leaved bittercress are specialities here.

The Dingle Valley Woods along the Teme have many similar characteristics to the other west Worcestershire Woods. Many ancient woodland sites on the south-west side of the Teme Valley near Shelsley Walsh have been replanted with poplars but old woods survive in the steep-sided dingles. Inaccessibility has allowed a number of special plants to survive here, including thin-spiked wood sedge, nettle-leaved bellflower and herb paris.

ABOVE: Coppice-with-standards in Hornhill Wood. (DMG) BELOW: New coppice growth under open standards of birch and other trees, Tiddesley Wood (DMG)

ABOVE: Derelict hazel coppice. (DMG) BELOW: Wyre Forest from the south, lying in the Dowles valley. (GHG)

LEFT: Shrawley Wood — five-year-old lime coppice attracts many summer-visiting warblers; (GHG) CENTRE: eleven year-old-lime coppice (same view as previous picture). As the coppice grows, garden warblers start to decline and blackcaps to increase; (GHG) RIGHT: tall lime trees grown from 'singled' old coppice form an attractive wood, Notice coppice sprouts from tree base. (GHG) BELOW: Teme valley dingle wood, rich in ferns, old trees and mosses. (GHG)

ABOVE: Lion Wood reserve with oak over holly and rowan on acid soils, (GHG) and BELOW: Brockhill Wood, near Lion Wood. which it resembled until cleared to create wood-pasture, so destroying much of interest. (GHG)

ABOVE LEFT: Nuthatches nest in holes in trees, often reducing the size of the entrance hole with mud. (AWC) RIGHT: Pied flycatcher, a typical bird of western oakwoods. (AWC) BELOW LEFT: Ramsons smell strongly of garlic and clothe wet woodland floors on lime soils. (DMG) CENTRE: Conifers shade out all the plants of ancient broad-leaved woodland in Tiddesley Wood. (GHG) RIGHT: Great butterfly orchid grows in woodland rides and smells weakly of cinnamon on summer's evenings. (GHG)

Golden Valley pool with yellow water lily and the Malverns beyond.
(DMG)

Lakes, Pools and Ponds

Lakes and pools are attractive and popular landscape features which are also rich in specialised wildlife. In Worcestershire most waters are basically similar, although they may appear different. The natural history interest of a pool depends on local structure, size, surrounding habitat and the degree of disturbance and pollution. People like 'messing about in boats' and many leisure activities impinge on water, causing conflict between humans and nature.

There are few large water bodies in Worcestershire, and no natural or man-made large lakes or meres over 32 ha (about 77 acres), and only 130 larger than 0.2 ha (about 0.5 acres), the arbitrary dividing line between small ponds and larger pools. A survey carried out by WNCT in 1977-78 recorded 300 water bodies in this category, occupying a total area of 377 ha (about 0.22% of the county). The following analysis is based on work by J. J. Day.

Although the number of sites seems quite high the total area is relatively small, making larger water bodies one of the county's rarer habitats. Despite their relative scarcity these water bodies are important wildlife resources. They are highly productive, with a thriving freshwater plant and invertebrate life and contain many special and unusual species, especially when the structure is varied with deeps and shallows, bordering emergent and marsh vegetation and associated woodland. The greater the variety of structure and habitat the greater the variety of species and population sizes. Fish and amphibians are likely to occur and the larger lakes support a wealth of bird life throughout the year, including ducks, grebes, rails, warblers and kingfishers.

All these waters are man-made and fall into three categories: ornamental pools; reservoirs (drinking water, canal feeders, mill pools and industrial pools), and pools associated with extraction industries (brick pits, gravel pits, subsidence pools caused by salt extraction). Most waters are in the north and central part of the county, especially near Kidderminster.

Of the 300 waters recorded, 225 were less than 1 ha, thirteen exceeded 5 ha and only five of these exceeded 10 ha, the largest being Upper Bittell Reservoir at 31.8 ha.

Size distribution of pools greater that 0.2 ha

Size (ha)	Number	% total number (ha)	area (ha)	% total area
0.2-0.5	154	51.3	48.2	12.8
0.5-1.0	71	23.7	47.0	12.4
1.0-2.5	40	13.3	58.5	15.5
2.5-5.0	22	7.3	76.7	20.3
5.0-10.0	8	2.7	58.7	15.6
10.0-20.0	3	1.0	32.4	8.6
20.0+ (max. 32ha)	2	0.7	55.8	14.8
Total	300		377.3	

The number and total area of pools over 0.2 ha have changed over the last 150 years, as shown by comparing the 1977-78 survey with the waters marked on first edition Ordnance Survey maps of around 1830, and with the first series of 1:25,000 maps of around 1916. The number of waters increased from 198 to 324 and the area from 240 to 388 ha. The increase in the number of pools between 1830 and 1916 of 111 was due to the creation of several large reservoirs, many ornamental pools, large pools for industrial purposes, with a few arising from extraction industries. The thirty pools lost in the same period were mostly associated with the disappearance of corn mills and silting of ornamental pools. Between 1916 and 1978 more extraction industry pools, amenity pools and reservoirs were formed and 36 other pools were lost, probably through silting, formation of marsh and drainage of more mill ponds.

Those waters classified as ornamental pools occupy a total area of about 166 ha and the class includes moats, new and old fish ponds and truly ornamental 'landscape' lakes, almost always associated with great houses and estates. Surprisingly, around 25% have been dug since 1965, often doubling as landscape features and modern fish ponds. Many of the new pools are still developing their aquatic, emergent and bank vegetation, and should prove to be excellent wildlife sites in the future. Some of the oldest pools are moats and fish ponds, and these usually have a well-developed aquatic but poor emergent vegetation due to their steep banks. About 80 pools were created primarily as landscape features, although some provided power for mills. Half were built before 1830 and the rest before 1900, and their existence owes much to the landscape architects of the eighteenth and nineteenth centuries: several Worcestershire estates were designed by Capability Brown, Croome being an outstanding example. There are several lakes on each of the Witley Court, Croome and Kyre estates which are often irregular in outline with island nesting sites for wildfowl. Many estate pools have shelving margins with extensive areas of emergent vegetation, reed beds and nearby woods and scrub, providing shelter throughout the year, and nesting sites for many birds. Three of the larger ornamental pools, Westwood, Pirton and Hewell, all in private ownership, are among the six top wildfowl sites in the county.

Westwood Great Pool is an important wildlife site, with a wide range of plants and birds. The lake and Westwood House were created around 1620 by Sir John 'Lusty' Packington as the centrepiece of an estate long since broken up. To form the lake he excavated a ravine (which can be seen between the boundary wall and the lake) in order to embank the water against the fall of the ground, and the lake is filled from a relatively small catchment area. It has suffered many changes in the last thirty years. It was drained in 1962, to remove

Great crested grebe — a bird of larger lakes. (WNCT)

coarse fish and re-filled slowly, so that in the late 1960s there were broad muddy margins and a large band of emergent vegetation. At the time the lake was relatively undisturbed and particularly rich in wildlife. In 1970 the water level was raised and the marginal band of exposed mud, (which had attracted waders), and of emergent vegetation decreased and it was used for private water-skiing. Later, attempts were made to develop angling and a trout fishery without much success and planning consents for future development for water-sports was not granted. During the period of increased disturbance and development the number of wintering wildfowl declined against the national trend but has risen again under more sympathetic ownership over the last five years.

Westwood Great Pool is generally less than two metres deep, increasing to 5.5 metres by the dam, and shelving to marsh on the opposite bank. Analysis has shown the water to be unusually nutrient-rich, probably due to fertilizer run-off from the adjacent arable land in the small catchment area around the lake. Nutrient-rich lakes are prone to algal blooms, when vast numbers of microscopic algae turn the water green and turbid, and high biological oxygen demand kills fish. Some blooms also produce toxic substances. Algal blooms occurred in 1983 and 1984 and further evidence of nutrient enrichment is the blanket of filamentous algae which covers parts of the lake bottom. Blooms occur in hot weather and can cause severe ecological damage.

About 150 species of birds have been recorded at Westwood, comprising 64 breeding species, including great crested grebe (up to ten pairs in the mid-1970s), little grebe (over 100 counted in the late 1960s) tufted duck, pochard, ruddy duck, Canada goose, mute swan, moorhen, coot, reed warbler, sedge warbler, reed bunting and many small birds associated with adjacent wetland and woodland habitats. Once, a surprise visitor was a Pallas's warbler, an exceptional inland record of this dazzling Siberian rarity. The lake is excellent for wintering water birds and records include Slavonian grebe, black-necked grebe, gadwall, pintail, wigeon, shoveler, scaup, goldeneye, long-tailed duck, common scoter, red-breasted merganser, goosander, smew and Bewick's swan. Before the adjacent parkland was ploughed in the early 1970s a wintering floock of up to 100 wigeon grazed the fields, accompanied by large numbers of coots and moorhens. Wigeon are now scarce visitors. Passage migrants include green sandpiper, wood sandpiper, spotted redshank, redshank and greenshank, all waders which feed and rest along muddy margins and are more frequent when water levels are low. As the only large sheet of water in the district, it attracts large roosts of gulls, including the occasional glaucous and Iceland gull during winter, and passage terns include black tern and white-winged black tern. Mediterranean gulls are now an almost annual feature of the gull roost.

Westwood Pool contains abundant submerged aquatic plants typical of eutrophic waters — Canadian pondweed, curled pondweed, fennel-leaved pondweed, spiked water milfoil, willow moss and stonewort. Plants with floating leaves are white water lily, amphibious bistort, and broad-leaved pondweed. The dominant marginal reed swamp plants are great and lesser pond sedges, common reed and great reedmace. In all, 41 species of fully aquatic plants have been found — 20% of the British species — and this number, together with the stonewort and willow moss, places Westwood among the best standing water sites in Britain. Interestingly, Upper Bittell reservoir contains 44 species of aquatic plants but less rich fringing vegetation. Over 220 species of flowering plant have been found at Westwood Pool.

Eight-stemmed waterwort — a rare plant found at Westwood Great Pool. (RMB)

Two extremely scarce water plants occur at Westwood, listed as Red Data Book Species — the eight-stamened waterwort and the ribbon-leaved water-plantain. The waterwort occurs at one other Worcestershire lake and is found nowhere else in England. It is a small, delicate, annual plant, with tiny unstalked flowers and four pink petals. It has a creeping habit and can form mats on firm wet mud at the water's edge where competition from other plants is minimum. The high nutrient levels at Westwood may encourage abundant growth of other species which could swamp the waterwort. The ribbon-leaved water-plantain is a perennial herb which can grow to one metre in height. It has a tuft of long, ribbon-like leaves and a much-branched flowering head with whorls of small, pale flowers each with three petals. It grows either completely submerged, as an emergent or on bare ground. In eutrophic waters it may not survive when coated with excessive algal growth. Seeds germinate in shallow water and the plant grows best where there is no competition from other plants. Apparently Westwood provides the unique combination of factors necessary for the plant to survive and it is now the only site for this plant in Great Britain. The plant is fully protected by law and may not be picked, collected or disturbed. No other British site contains both of these rare species, which may have arrived on the feet of migrating waterfowl.

The other large and small ornamental pools in Worcestershire contain similar mixtures of plants and animals to Westwood, generally in less abundance and variety, without such rare plants, but with their own particular attributes. Pirton Pool, for example, is regularly visited by wintering Bewick swans, which breed in Arctic Russia, wearing colour rings from Slimbridge. It is notable for its gatherings of gadwall, a local species in the county.

There are about 130 reservoirs in the county, occupying 159 ha, and of four main types: agricultural irrigation supply pools; mill pools; drinking water reservoirs and canal feeders. The agricultural pools are either of recent origin in areas of commercial orchards and market gardening, or large farm and village ponds, with the characteristics of ponds. Some of the newer agricultural ponds are rubber-lined and usually devoid of wildlife; however, one in the Vale of Evesham must have been placed on or near an old pool site as it was colonised by large numbers of newts, frogs and toads, which blocked the pipes and pumping system — many hundreds were transferred to other waters.

Before the extensive use of coal and oil, and through much of the nineteenth century, water-driven mills were common features along streams and rivers. Streams were dammed to create mill pools to drive waterwheels and in Worcestershire the 1977-78 survey recorded 58 mill pools, mostly at old corn mills. Industrial demands led to larger storage pools to provide more power, the biggest still in existence being Hurcott Pool. Water-powered mills were used for paper-making and associated with nail, chain and tool-making (especially scythes), mainly in the northern part of the county. These mills are a special feature of the tributaries of the Stour where there are several series of pools, often four or five in a valley. These are now often interspersed with willow and alder carr woodland or marsh, extending up the valley from the pool. The fast-flowing feeder streams carry much silt which is deposited at the top of the pools, creating delta-like areas where marsh vegetation and eventually woodland develops. These wet woods are unusual in the county and typical ground plants are large bittercress and greater tussock sedge. The larger pools are good for birds, including breeding mallard, moorhen, coot and great-crested and little grebes.

There are four canal feeder reservoirs, totalling 50 ha, all serving the Worcester-Birmingham canal. The three Bittell reservoirs probably comprise one of the best-watched bird sites in Worcestershire, and two are SSSI. Upper Bittell is 40 ha in extent and the largest water body in the county. The lower 23 ha reservoir is smaller but has a well-developed emergent bordering vegetation, and a rich bird population. It is divided in two by a road which creates a smaller pool to the north between the two main reservoirs. A stream drains into this pool through an alder copse which attracts siskins even in poor years. Bittell is outstanding at all seasons, especially when water levels are low and a muddy shoreline exposed. Many waders occur on passage, both in autumn and spring: greenshanks, spotted redshanks, ruffs and common sandpipers are regular visitors and little-ringed plovers stay to breed if conditions are right. Up to fifty great-crested grebes have been recorded in winter, several pairs remaining to breed, and Slavonian, black-necked and red-necked grebes appear regularly enough to raise the ornithological blood-pressure. Diving ducks are seen regularly on Upper Bittell where the water is deeper and there is an ample supply of swan mussels: eiders and velvet scoter have occurred, though up to 100 pochard and 160 tufted duck are more usual. Watching here can be a bit hit-and-miss: be sure to avoid a day when the sailing club is out in force. Much of the Upper Bittell shoreline is flakey shale of remarkably adhesive properties, deterring all but the most intrepid of naturalists but popular with the inconspicuous mudwort, which possibly has its largest Midland population here. As Harrison and others note in the *Birds of the West Midlands,* Bittell has most of the county's records for seabirds. Sandwich and little terns are occasional visitors with larger numbers of 'commic' and black terns, the latter arriving in the wake of easterly winds during passage times — 100 on 5 August 1954.

There is a rich variety of habitats round the reservoirs, with extensive areas of fringing marsh, scrub and small alder woods. 81 bird species have been seen in winter, as many as at Westwood Pool, but there are fewer breeding and passage species. In recent years the reservoirs have been visited by a small flock of wild barnacle geese and there has been much speculation on their Arctic origins — Greenland, Svarlbard or Russia?

Twenty five pools, total area 45 ha, supply drinking water or are industrial reservoirs. Most are poor wildlife habitats as they have concreted banks and are free of emergent vegetation. In the north, Bartley Reservoir, which straddles the Worcestershire-West Midlands border, is quite good for wildfowl and has a large gull roost. Trimpley, by the Severn north of Bewdley, occasionally holds interesting waterfowl and cormorants.

The county has 39 pools, totalling 53 ha, formed by extraction industries, or as settling tanks associated with waterworks and sugar-beet washing. They include clay pits, sand and gravel workings, quarries and subsidence pools associated with salt.

In the nineteenth and early twentieth centuries many pits were excavated by the Severn between Tewkesbury and Stourport to provide clay for the manufacture of bricks and tiles. The disused, impervious pits filled with water and have developed into wet willow or alder woods, or marshland. Several still contain small ponds, often similar in character to large farm ponds, and four contain larger open waters, including large pits at Upton-on-Severn which have been connected to the Severn and are now used as a marina. Probably the most important brick pits are at Grimley, with their marshland habitats. The Grimley pools are used by breeding mallard, coot, moorhen, tufted duck, Canada geese and mute swans. In winter the pits are often surrounded by flood waters and then attract many more species, including white-fronted geese, pochard, shoveler, up to 100 wigeon, 200 teal and 100 mallard. Lesser black-backed gulls, herring gulls, black-headed gulls and lapwings are also numerous on the flood waters and their margins. During cold weather, when many lakes and pools freeze, the river next to the pits is occasionally used by wintering goldeneye, scaup and long-tailed ducks.

Sand and gravel pits are an active feature of Worcestershire, and are based on the post-glacial river terraces of the Severn and Avon valleys and on the Triassic sandstones north of Bromsgrove. The northern sand pits are mainly dry. The 1977-78 survey recorded 136 ha of sandpits and 109 ha of gravel pits with 51 ha from both recorded as disused. 26 ha of the disused pits were being filled with rubbish. There are important sand and gravel workings between Grimley and Holt on the west bank of the Severn north of Worcester, in the south near Bredon, and at the smaller pits near Lower Moor and Kemerton.

The Grimley-Holt pits have been in use for many years and vast quantities of gravel have been extracted. Most of the land in this district is returned to agriculture after extraction so their abundant and varied wildlife is of a transitory nature. The pools tend to be fairly shallow and often burdened with silt from washing plants. They are rapidly invaded by emergent vegetation, particularly great reedmace and are often good for birds. Small islands are used by nesting Canada geese and mute swans, and mallard and moorhens may breed where there is some cover. The migrant waders — common sandpiper, green sandpiper, greenshank and redshank — are attracted to pools with shallow margins and large flocks of moulting lapwings may gather in late summer, many from continental Europe; a Grimley-ringed bird was subsequently found in Czechoslovakia. In the past a large (100-200) post-breeding moulting flock of Canada geese gathered at the pits. Canada geese have increased markedly throughout the county during the last twenty years, with pairs nesting at many waters, usually on islands. At times they become an agricultural pest where they gather into winter flocks and graze and trample both cereal crops and grassland near pools.

The varied structure of gravel pits and the extraction and washing processes encourage a wide range of wet and dry habitats to develop, with a large variety of plants and their associated insects. Exposed and dry areas are invaded by weeds of many species and their seed crop provides an autumn and winter food supply for large flocks of finches and sparrows. Almost any 'agricultural weed' can be found, and docks, thistles, and fat hen are often common. Thorn apple regularly occurs in the dry parts of Worcestershire pits. This is a poisonous plant, originating from America and now naturalised throughout England. It may grow two metres high and has large, toothed leaves and trumpet-shaped, white or purple flowers, which produce a prickly ovoid seed capsule up to five cms diameter, containing many black seeds. All parts of the plant are poisonous, containing a cocktail of alkaloids, and its conkery fruits luminesce when soaked in surgical spirit.

The great increase of gravel workings in Britain since 1945 has been responsible for the steady increase and spread of little-ringed plovers from south-eastern England north-westwards. The species reached Worcestershire in the 1970s and has bred at most gravel pits with areas of level open ground and pools ever since.

Wetter areas provide feeding and breeding sites for insectivorous wetland birds such as reed warblers, sedge warblers and reed buntings. The tall, emergent vegetation, willow scrub and reed beds are also important swallow, sand martin and wagtail roosts. Colonies of sand martins nest in near vertical 'cliff' faces and more are found in pits than along the rivers.

In times past planning consent for gravel-working usually required the pits to be returned to agriculture, although some were used as rubbish tips. Some have remained wet; for example, the large pool at Upton Warren is used as a sailing lake, and it, and the associated willow scrub and reed bed, form part of a larger SSSI and nature reserve. In more recent times development of gravel extraction has included permanent formation of wetland, for example at Bredons Hardwick by the Avon, and parts of pits near Kemerton which will eventually be managed as nature reserves. Recently, planning consent for extraction south of Worcester, near Clifton, has included an obligation to create wetland habitats and a wildlife reserve.

WNCT has one tiny gravel pit reserve, at Milestone Ground just outside Broadway, situated on stream-washed fan gravels, derived from the oolitic limestone of the nearby

Moorhen among mare's-tail — a bird of pools, streams and rivers and an uncommon plant of base-rich waters. (RMB)

Cotswold escarpment. This tiny two-acre site is surprisingly rich in wildlife and brings wetland species into a dry part of the county. Within a small area there are willow scrub, ponds, a small bed of reedmace, a marsh, dry banks and tall rank vegetation on wet and dry ground, bare ground and pockets of grassland. The pools contain masses of mare's-tail, a waterweed uncommon in the county and usually found in limey conditions, a water crowfoot *Ranunculus trichophyllus* and abundant insect life. Pool fringe plants include both pink and blue water-speedwells and the marshy area contains common spike-rush, jointed rush, sharp flowered rush and hard rush. The dry areas carry a ruderal flora (ie: those plants which first invade disturbed ground), including wild carrot, hoary ragwort, bristly ox-tongue, wild lettuce and narrow-leaved pepperwort. Common darter dragonflies occur and grassland butterflies such as marbled white, small skipper, small copper, meadow brown and ringlets. Moorhens, mallards, little grebes and sedge warblers breed. The pit is also good for a wide range of insects. Last but not least it contains slime-mould fungus, *Didymium applanatum*, never before described in Britain and first described in Holland in 1972 where it was found on the sandy floor of a polder being drained and reclaimed — a habitat much like Broadway Gravel Pit!

Salt extraction has caused subsidence and the formation of pools at Upton Warren and Oakley (Pulley Farm), all of which have developed into important natural history sites and are scheduled as SSSIs. The complex of pools at Upton Warren is a WNCT reserve and an ornithological site of regional importance. The site of the main Moors Pool was clearly farmland as shown on an estate map of 1868. In 1968, 100 years later, when the pool was purchased by Christopher Cadbury and the reserve was established, old fence lines ran through the water, which had flooded a public footpath. In the last five years the pool has sadly declined, probably due to changes in water chemistry or pollution, as yet not understood, and the aquatic plants, invertebrate and bird populations have all become poorer. Tufted duck, ruddy duck and little grebe, which bred for many years, have gone, and those great-crested grebes which attempt to breed fail to rear chicks. Wintering ducks have decreased, especially shoveler. Extensive areas of amphibious bistort have died. Fortunately the smaller car-park pool is still in good health, with breeding mallard, gadwall, ruddy duck, coot, moorhen and water rail, and several species of dragon and damzelfly. The Flash Pools to the south are a series of shallow pools fed by salt springs and of varying salinity, with a small saltmarsh. The salt depresses plant growth and there are always bare areas on the banks and artificial islands which attract waders, including hundreds of lapwings and breeding little-ringed plovers. The fields beyond the Flashes are short-turfed and grazed and frequented by a flock of wintering curlew. The land round the Moors Pool contains willow scrub and marsh, and alders and pollarded willows line the river Salwarpe which passes between the Moors and the County Council's education reserve based on old gravel pits.

Bird-watching hides have been built at both the Moors Pool and the Flash Pools and Upton Warren is the most intensively bird-watched site in Worcestershire. The bird list continues to grow, having topped 210, and many watchers believe that at least one of practically any bird species will, eventually, arrive at Upton Warren. Bittern, hen harrier, Iceland gull, Slavonian grebe, velvet scoter, marsh warbler, Savi's warbler, golden oriole, bluethroat, alpine swift, purple heron, blue-winged teal, little gull, Sabine's gull, Mediterranean gull, Leach's petrel, gannet, long-tailed skua, Wilson's phalarope, Caspian tern, yellow-browed warbler, Cetti's warbler, rustic bunting, smew, scaup have all been seen, together with almost every wader you can think of, including America's pectoral and least sandpipers. The site is interesting throughout the year and for those who are bored by birds there is a long plant list, as well as abundant insects, dragonflies and twenty species of butterfly, including Camberwell beauty. Many birds are found at Upton Warren because it is the only large, open, wetland habitat for miles around and so provides a haven for migrating wetland species. Locally, waterfowl often move between Upton and Westwood, Great Pool and Oakley Pool which are both to the south, unless they are displaced by freezing weather, when they either leave the district or congregate on the river Severn north of Worcester. In a few parts of the county, small lakes fed by underground water remain open in freezing weather and attract large numbers of duck.

Upton Warren has been enhanced still further by a feeding station, which attracts all common woodland and garden birds as well as bramblings and water rails.

North American ruddy duck, a stiff-tail diving duck, was imported to Slimbridge in 1948 and bred successfully in captivity. Between 1952 and 1973 at least 70 juveniles escaped and, in the cold winter of 1962-63, all the free-flying birds left Slimbridge. At least 50 pairs were nesting in the wild by 1975, mainly in the West Midlands. The first breeding birds were found in Worcestershire in 1971 at Upton Warren and Westwood Great Pool. Numbers have increased steadily ever since and the species breeds at many pools in the county. It is a rather secretive duck, remaining among emergent vegetation, and it has a long breeding season, with young produced from May to October. It feeds mainly by diving and straining ooze to collect invertebrates and plant material. Interestingly, ruddy ducks now occupy an ecological niche previously empty in Europe and their spread is likely to continue.

At Wilden Marsh the British Sugar Corporation discharges sugar-beet washings into large ponds which gradually fill with silt. During the silting process the muddy edges of the pools are attractive to waders such as green sandpiper, common sandpiper and spotted redshank, and the water attracts ducks. A great many swallows, house martins, sand martins and swifts hawk for insects over these pools during the summer.

The wildlife of many of the large pools and lakes in Worcestershire is under threat of damage or destruction. The most serious factors are pollution, silting and disturbance. Pollution may arise from sewage outflows, effluents from pig farms and cattle pens, run-off of agricultural fertilizers, spray drift and dumping. Eutrophication by agricultural fertilizers is a serious problem although contamination by herbicides or pesticides may be even more serious. Silting is a natural process which occurs when the silt burden from fast-flowing streams is deposited when water flow slows, as in a mill pond, and mill ponds in the county have been completely filled by silt. They may then turn into marsh or wet woodland.

Leisure activites cause excessive disturbance at times, by scaring away sensitive birds, destroying emergent vegetation, and by associated pollution. Boating, water ski-ing, wind surfing and sailing can all cause damage but much depends on the intensity of these activities. Ideally some waters should be free of these activities or large parts of them should be set aside as disturbance-free zones. Angling is often compatible with wildlife (except for the fish) as anglers have a considerable interest in maintaining water purity and fish stocks, and some pools made especially for fishing provide habitat for a wider range of wildlife. Discarded fishing ines and hooks kill many birds and lead shot used to weight line kills mute swans.

A careful study of the first series OS 1:25,000 maps of Worcestershire, surveyed around 1916, has shown that there were then 7,653 ponds less than 0.2 ha in extent, covering a total of 472 ha, greater than the total area occupied by larger waters. A similar comparative study of OS maps surveyed around 1971 (available for 70% of the county) revealed 3,593 ponds in an area which contained 5,531 in 1916. An estimate made from this data for the whole county suggests there were about 5,000 ponds in the early 1970s, covering a total area of about 310 ha. Loss has certainly continued since then and by the time of WNCT's 1977-78 survey there were probably about 4,550 ponds in the county, a loss of about 3,000 ponds since the beginning of the century, about 25-32%. Further losses have continued.

In Worcestershire the vast majority of small ponds are man-made (usually by clay puddling) farm ponds for watering stock. Others were constructed as fish or duck ponds and in parts of the county developed in small marl, clay or sand pits. Farmyard and village ponds provided drinking water for stock and a place for rearing ducks and geese. The number of ponds increased during the nineteenth century during the Enclosures to provide watering places in the new fields, and most of these remained in use until World War II. The subsequent decline in the numbers of ponds, and of their water quality, followed changes in agricultural practice: larger fields, increased arable farming in traditional livestock rearing areas, decrease of grazing stock, appreciation of the role of ponds in the spread of animal diseases (for example, liver fluke with a life cycle involving water snails), and the ready availability of piped water. In recent times a few new ponds have been established for irrigation, and ponds have become increasing popular in gardens.

Unless cleared out regularly, ponds disappear through silting, marsh formation and transition to drier habitats. Willows planted near ponds give shade which suppresses the growth of aquatic plants and provide poles from regular pollarding. Leaf and twig fall aids silting and old dry pond sites can be seen next to groups of willow trees, although willows are often removed and the pool disappears into a level field. All these features can be seen in Worcestershire and 32% of 1,500 ponds examined in the 1977-78 survey were in an advanced state of silting with little wildlife, while another 10% had developed into marshes. Many farm ponds have suffered a worse fate, being used as rubbish dumps, sometimes for discarded chemical containers, or suffering from crop spray drift or fertilizer contamination. In the latter case they suffer repeated algal blooms and freshwater life dies. It is a sad experience to re-visit a pond once full of waterweed, invertebrates, amphibians and small fish, and to find instead a turbid, stinking mess: such experiences have become all too common.

The 1916 OS maps show that then the majority of Worcestershire's ponds lay on the Keuper Marls in a band crossing the centre of the county from north-east to south-west. This distribution was related to small field size and the high proportion of pastureland in this belt, but the total is inflated above likely need by the enormous number of marl pits in the north-west, rising to as many as ten in one field. These pits were used mainly for digging out marl to spread over the field as a fertilizer, although some provided clay for brick-making. In 1916 ponds were next most abundant on the Lias clays lying south-east of the Keuper Marl with smaller numbers in the market gardening and orchard region of the Vale of Evesham. Fewer ponds were found in the north and west of the county, where there were plenty of streams to provide drinking water for stock, while the free-draining Triassic sandstones were mainly arable land with few ponds. Present day ponds follow much the same distribution as in 1916, with a greater proportion of pond loss on the Lias clays than elsewhere. By 1978 the Keuper Marls had lost about one-third of their ponds, which is less than might have been expected, as there had been a large increase in arable land. Perhaps marl pits survive because they are deep and so not easily ploughed-out and filled-in.

The wildlife in and around a pond, assuming it is in a healthy condition, depends on the amount of shade from trees. Light is essential for good growth of normal algal freshwater plankton on which much else depends; in fact, all the complicated interlinking strands of a pond's food-web depend on the large plants, the phytoplankton and the filamentous algae. Their growth depends on photosynthesis, the process in which light energy, carbon dioxide and water manufactures sugars necessary for green plants to exist. Partial shading does, however, increase the habitat range in the pond and helps keep it from over-heating in summer. Some trampling by stock of muddy margins also increases habitat range. Size is also important: there is generally a wider range of flora and fauna in bigger ponds. Coot, for example, are rarely found on ponds smaller than 1,000 square metres, whereas moorhens nest on the smallest ponds. In the 1977-78 survey, 58% of ponds recorded were less than 500 square metres, 23% between 500 and 1,000, 14% between 1,000 and 1,500 and 5% between 1,500 and 2,000 square metres.

Nowadays only about 2,700 Worcestershire ponds, occupying about 169 ha, are still of natural history interest, a remarkable decrease from the 7,653 ponds covering 472 ha marked on maps at the turn of the century. The surviving good-quality ponds are mainly in the south and south-west of the county. The marl pit ponds in the north-east usually contain only a few species and are heavily shaded with steeply wooded sides. The dry pits may be good for badgers.

A good-quality living pond is an active place, albeit often at a microscopic or slightly larger level. The plants are the engine which drives the pond's web of food chains. The most numerous organisms are likely to be microscopic bacteria and fungi living on, and decomposing, the detritus of dead plant remains in the bottom of the pond. Both detritus and bacteria are eaten by larger single-celled protozoa: the familiar amoebae and ciliates of school biology textbooks. Next in abundance are microscopic diatoms (which are plants) with their intricately sculptured silicate cell-walls and the free-swimming plants of the phytoplankton.

These are eaten by small crustaceans, the most familiar being *Daphnia*, the so-called water-flea, because its shape is similar to a flea, and *Cyclops*. The small crustaceans and the phytoplankton are eaten by water snails, tadpoles, water beetles, dragonfly and mayfly nymphs, water boatmen, fish-fry and many others. The filamentous algae (for example, *Spirogyra*) are eaten by water snails, herbivorous beetles, and tadpoles. Then there are the carnivorous water beetles, dragonfly nymphs and water-boatmen which prey on water snails, each other and tadpoles. At the top of the food-chains are fish, usually three-spined

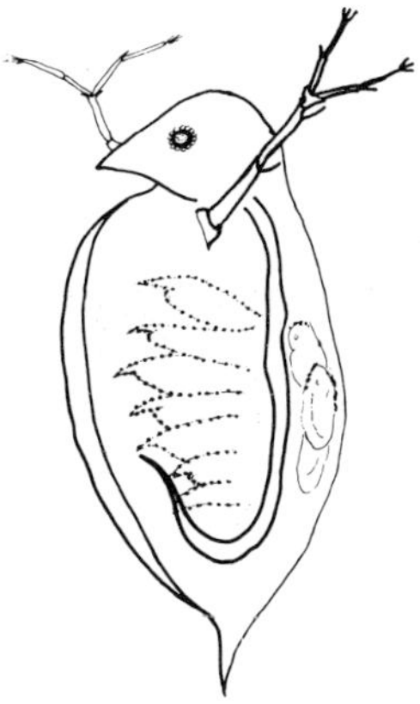
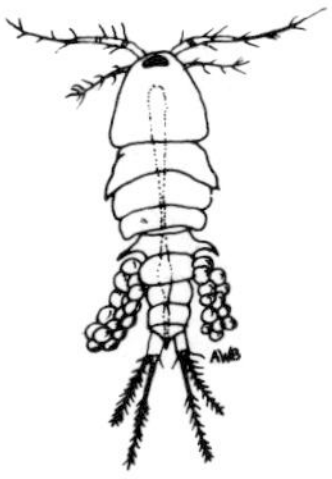
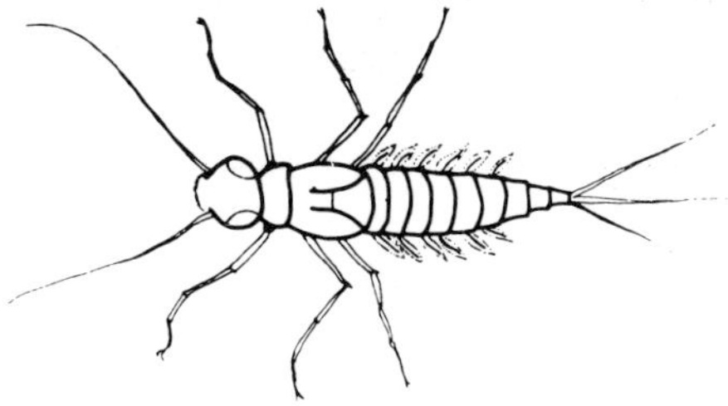

LEFT to RIGHT: Pond life — *Daphnia, Cyclops,* and mayfly larva. (WNCT)

sticklebacks which prey on the insect nymphs and adults, and tadpoles. There are also the caterpillars of several species of china-mark moths which eat water weeds, and water spiders which spin bell-shaped webs under water which are then filled with air dragged down by the spider which feeds by catching passing animals and taking them to the bell to eat.

The big hawker dragonflies are among the most spectacular of insects, mottled brown, green and blue, and the southern and brown aeshna are quite common around Worcestershire's ponds, lakes and rivers. The chasers *Libellula depressa* and *Libellula quadrimaculata* are splendid large insects with broad blue or brown bodies and typical pond species, often hawking and hunting round ponds for long periods. The common darter is another common pond species, often lifting the tip of the abdomen in a typical stance when it is perched. Individuals usually have a favourite perch to which they return again and again. Several of the blue and black damzelflies occur regularly at ponds: the blue-tailed damzelfly, the common blue damzelfly and the azure damzelfly.

Ponds also contain many worms of different groups, from flat scavengers to leeches; innumerable insect and midge larvae, many tiny crustaceans, different flies, fish and amphibians. The latter are the newts, frogs and toads. Common frogs have declined with the disappearance of ponds and, although generally well-distributed, they are less abundant. In recent years various 'Frog Watch' campaigns through the RSNC's WATCH club for children have brought in lots of records. Frogs can be abundant at some traditional ponds, reaching plague-like levels! Gardens of new houses built on old frog sites may swarm with frogs in late winter and early spring, when the adult frogs return to pair and mate, much to the astonishment of new owners. Many buckets loads of frogs were transferred from a Worcester garden to start colonies elsewhere, and to help the desperate owner deal with frogs in everything, including the front room. New garden ponds are partly compensating for loss of ponds in the countryside. Frogs of course lay their eggs (frog spawn) in ponds and these hatch to produce vast numbers of tadpoles, a few of which survive to develop into froglets, which leave the pond to live and hunt insects in damp places in the countryside, before returning to their natal pond to breed. This is the cycle which builds up large frog colonies at particular ponds.

Toads are found in many wetland situations and prefer to breed in deep ponds. They return to mate at traditional ponds, often from long distances. Strangely, they seem to cross roads to reach their ponds far more often than frogs, where they get killed by traffic. A considerable effort has been made to mark roads to warn motorists of toad crossing places and to try to reduce mortality by picking up the toads at night as they cross the road. Huge numbers of toads can gather in a traditional pond, the males forming balls round the unfortunate female as she lays eggs in strands on submerged vegetation. Both frogs and toads eat insects and slugs, and toads under the potting shed make an excellent slug-control system.

ABOVE: *Aeshna mixta,* an uncommon dragonfly. (WNCT) BELOW: Male great-crested newt — now rare in Worcester's pools. (WNCT)

Grass snakes may be found in ponds, especially if frogs are about, as they are a favourite snake food, although a wide range of prey is taken. They are most frequent in Worcestershire's river valleys, especially in marshy areaș. Grass snakes lay eggs in heaps of rotting vegetation and manure. Compost heaps may be used, so a garden with a frog pond, lots of damp tall vegetation and a compost heap may well attract these snakes.

Smooth newts are common in wetlands and rural ponds. They and their immature aquatic newt-poles eat most pond animals, including frog tadpoles, so the two are rarely found in the same pond. Great crested newts are large, black and scarce. Numbers have decreased so much that they are fully protected under the Wildlife and Countryside Act. Fortunately they have been found at a number of Worcestershire's ponds, particularly on the Keuper Marl, and the county may hold 200 breeding sites.

The commonest plants in Worcestershire's good quality ponds are the duckweeds: *Lemma minor* is widespread but *L. triscula, L. polrhiza,* and *L. gibba* are only found on a few sites. Water milfoil and the two hornworts are widespread but local. Two water-crowfoots are widespread — *Ranunculus peltatus* and *R. hederaceus* — as are starworts, Canadian pondweed, and pondweeds *Potamogeton crispus* and *P. natans*. Emergent and marsh plants are bur-reed, reed mace, greater and lesser pond sedges; false fox and hairy sedges; soft, hard, jointed and toad rushes; floating sweet grass, reed sweet grass, marsh foxtail, yellow flag, blue water-speedwell, gipsywort, purple loosestrife, common spike-rush, marsh bedstraw, great hairy willow herb, brooklime, amphibious bistort, water pepper, celery-leaved buttercup and watercress.

Several ponds hold plants rare in the county, especially great spearwort, marsh cinquefoil, and bogbean. The floating water fern, *Azolla,* which is gradually spreading throughout the county, is found on small ponds. Many birds use ponds as drinking places and for insect food. Feeding herons visit ponds in search of fish, frogs, small mammals and young birds and are often the early-morning visitors to garden ponds which cause goldfish to disappear.

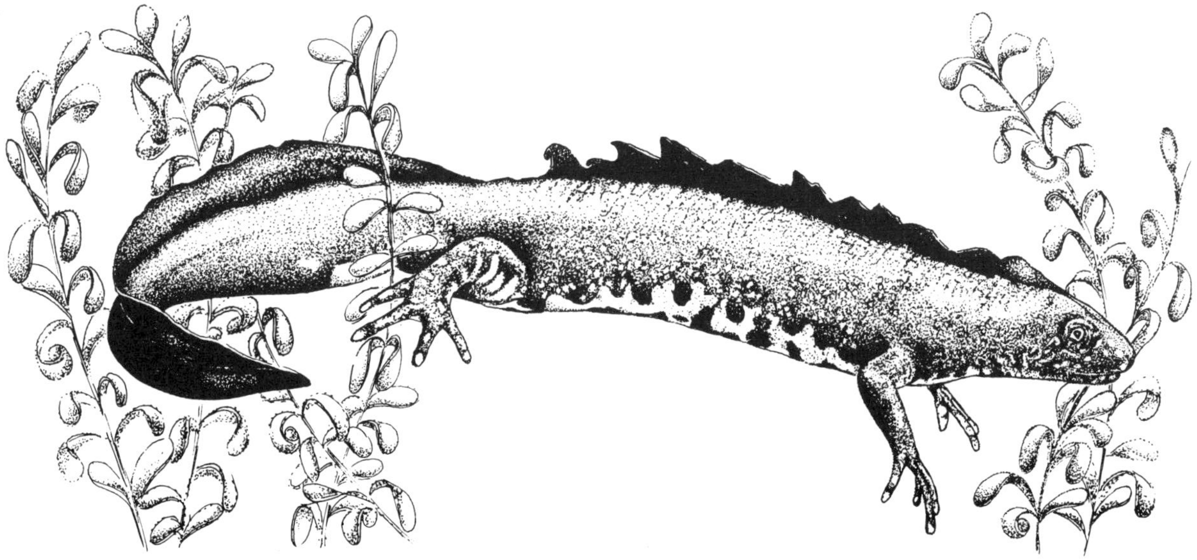

ABOVE: Pirton Lake, with artificial island for ducks and nesting Canada geese. (DMG) CENTRE: Westwood Great Pool — frozen solid but more usually a lake rich in birds and uncommon plants. (DMG) BELOW: Old brick pit pool near Ripple — full of fish and abundant water lilies. (DMG)

ABOVE: Luxuriant vegetation at a disused gravel pit near Beckford, used by many birds. (DMG) LEFT: Bird island in the salty flashpools at Upton Warren reserve, visited by many unusual waders. (DMG) RIGHT: Over-neutrified farmpond full of smothering algae. (DMG)

LEFT: Pill sedge, sorrel and ling colonising burnt ground at Devil's Spittleful. RIGHT: Top to bottom: Green tiger beetle, minotaur beetle, male emperor moth and heath orb spider, *Aranus quadratus*. (All DH)

Hartlebury Common seen through a summer heat haze. (DMG)

Heathland

'A wild space of land consisting of two distinct types — a sandy waste at the top of a steep bluff and a moor-like expanse at the bottom of it'. So reads the *Victoria County History of Worcestershire,* describing Hartlebury Common, near Stourport-on-Severn, with more than a hint of Hardy's Wessex. The modern visitor may find that the area falls a little short of this extravagant blurb — Worcestershire is not widely known for its heathland, but even so there are a few choice pockets remaining.

All lowland heath in the county is restricted to the north and north-west, where it overlies the Triassic sandstone formed two million years ago from deposition of riverine silt and sand. As you travel towards the Kidderminster area from the south, outcrops of bright red sandstone appear together with a characteristic flora; some fine examples draped with great woodrush can be seen along the Kidderminster ring road. A huge area stretching from Stourport across to Bewdley and north along the Stour valley was once a solid tract of heathland which still betrays its origins. Derelict land and road verges sprout saplings of birch and gorse as the original habitat tries to re-assert itself.

Of course, heath is very much a transitional habitat and, left to its own devices, would soon become woodland. How then do we define heathland? It is generally characterised by poor, usually acidic soils and has a specialist flora of which the heathers are a strong component. All Worcestershire heaths are predominantly dry and there is only a solitary example of acid bog left in the county. Fortunately, it lies on one of the two principal sites, both of which are managed to conserve their status.

Anyone involved with the management of heath will confirm that it is like painting the Forth Bridge — it never ends! A constant watch must be kept for colonising birch and oak, the bane of heathland conservationists everywhere. An excellent example of current practice can be seen at the Devil's Spittleful, a 62-acre nature reserve owned by WNCT on the western fringe of Kidderminster. Together with the adjacent Rifle Range reserve, the area is the largest stretch of heathland in Worcestershire. Its survival has always been tenuous; suburbs encroach from the north and lions roar in the neighbouring safari park. WNCT, in liaison with Wyre Forest District Council (which owns the Rifle Range) and the County Council have reduced bracken, birch and hawthorn, promoting fresh heather growth.

Approaching the Spittleful Reserve (sometimes called Spadeful) you notice first the 'spadeful' itself, a nub of worn sandstone crowned by Scots pines. This is a mass of rock reputedly flung by the devil to block the course of the river Severn. Now it stands proud of a tangle of re-generating birch and oak while all around the heathland is flat, rising to the north to form a natural amphitheatre.

The dominant plant species typical of lowland heath — gorse, ling, broom, bracken, oak, silver birch and hawthorn — constantly encroach, threatening the less robust and more exacting plants, which reward the diligent searcher. Several species benefit from the regular heather burning conducted by the conservationists or, less charitably, by careless smokers. Grey hair grass is the most surprising of these, its distinctive bluish tufts dotting the charred ground near the railway line. It presence here is suspicious for it is generally a plant of coastal sand dunes in East Anglia and Scotland. However, it does grow on Kinver Edge, five miles to the north-west.

Trains are certainly responsible for the meadow saxifrage, which rims the clinker with a tidemark of white, stretching from the reserve into Bewdley. Now a declining plant in Worcestershire, this lover of the sandstone often signals the occurrence of more recondite rarities. These are not easily found. Only close scrutiny of the open turf will reveal the tiny purple blooms of spring vetch or the spiky cushions of annual knawel. Pathsides and bare sand are ideal habitats for early forget-me-not and shepherd's cress; heat-seaking annuals which flower in spring, shrivelling in the heat of the summer sun.

Heat is an important consideration for many heathland plants. Some have reduced the effects of exposure to the sun by growing narrow leaves to limit evaporation. The heathers have refined this technique and have tough, waxy coatings to their foliage. Other species use heat to their advantage. Stand next to a broom thicket on a hot July afternoon and listen to the cannonade of pods splitting to catapult the seeds several feet away. It also pays to be opportunist. Some Spittleful rarities are annuals which cannot tolerate competition. Tower mustard is a spectacular plant which flourishes in the vicinity of rabbit warrens. Its impressive spires of creamy flowers, followed by sickle-shaped seed pods, only appear when the ground is disturbed, so that finding this capricious crucifer is cause for botanical celebration. Bur chervil, a delicate umbellifer, has similar requirements and is more frequent in this corner of the county than in other parts of Worcestershire.

Fringing the heath are a number of small, acidic meadows with a few specialities. Wild pansy or heartsese, the forbear of the gaudy garden favourite, produces its bicolour blooms on the top of well-drained anthills, where it can survive without competition. A much rarer plant lurks in the pony-cropped turf of one field. Each year a few spathes of moonwort appear at this, its best county site. This diminutive fern was once widespread, but has been dramatically reduced along with its favourite habitat of old meadows. It is, however, remarkably inconspicuous and has probably gone unnoticed in a few places. Worcestershire's second site was discovered nearby in 1990, during fieldwork for a new county flora.

Moonwort would not be helped by the fires which range regularly across the reserve. Their long-term value is undisputed, but in the short-term they destroy a large number of invertebrates for which heathland is renowned. Autumn sees the appearance of orb spiders on the heather tops, glowing orange and chocolate like luscious sweets. With them you can often find *Mangora acalypha,* a small, richly-coloured species with a white-bordered oakleaf pattern on its abdomen. Down in the undergrowth are the grey funnels of the labyrinth spider, a large and reclusive creature which guards its eggs, the female dying on 'sentry duty'.

Nightjar hawking for dusk-flying insects.

Spiders are not the only predators. In March and April green tiger beetles glint along the paths, flying off when disturbed. Their larvae live in burrows, sealing the entrace with their flat heads. Food walking literally overhead is seized in a backward flip, and sucked dry. Another fast-moving predator is the bug *Alydus calcaratus*. Slim and black with an orange-banded abdomen, the adult may be a mimic of the sand wasps which frequent the heath. The bug practises double-deception, for its larvae closely resembles an ant.

Devil's Spittleful is well-known to lepidopterists for its rich moth fauna. A few, such as the emperor, are diurnal, but others are more elusive. Waved blacks were once thought to be restricted to the Bagshot heaths of Surrey, but have since been found in a number of localities in the Severn valley. The caterpillars feed on bracket fungi and careful examination of birch polypores may reveal their presence. Other notable moths include the true lover's knot, heath rustic, angle-striped sallow, and lime hawk which feeds locally on birch. About thirty species of butterfly have been recorded. In these days of declining numbers it is heartening to report that silver-washed fritillaries, green hairsteaks and ringlets are increasing, though on the debit side pearl-bordered and small pearl-bordered fritillaries are rarely seen.

Reptiles and amphibians are not as common as may be expected. Slow-worms and common lizards are occasionally found, though adders present an enigma of distribution. These snakes occur in the nearby Habberley valley, once contiguous with the Spittleful, and at Kinver and in the Wyre Forest, but not here or on Hartlebury Common. Their absence may reflect past land use — certainly fragmentation of heathland prevents any chance of recolonisation.

A painful experience awaits the modern ornithologist who looks back at old records of Worcestershire's heathland birds. The roll-call of species is a depressing list of losses, most of which vanished earlier in the twentieth century. Red-backed shrikes and woodlarks bred regularly at the Spittleful, and the mechanical reel of the nightjar or 'eve churr', was a feature of hot summer nights. Gone too, are stonechats and whinchats as breeding birds, though both still turn up on passage. Destruction of habitat played a large part in their disappearance, but it was clearly not the sole factor. Nightjars, for example, have staged occasional comebacks and one in 1977 held territory in the adjacent safari park, competing with the roaring lions.

Hardier birds remain and tree pipits and whitethroats are conspicuous visitors. Garden warblers are especially fond of hawthorn and broom thickets, with up to 15 pairs breeding in good years. In winter these migrants are replaced by fussing tit parties in the undergrowth and redpolls and siskins in the birches.

South-east of Stourport is Hartlebury Common, Worcestershire's richest and best documented heath. To the first-time visitor it appears as an open mass of sand and heather, ostensibly devoid of wildlife, but such appearances are deceptive. The Common is the last enclave in the county for a surprising number of species and owes this distinction to its geological and sociological past.

There are few inland sand dune systems in the British Isles, and Hartlebury Common presents an unfamiliar landscape in an otherwise lush and agricultural county. The bedrock underlying the Common derives from the Triassic period, which is divided in Britain into the Keuper and Bunter series of rocks. When these rocks were formed, Britain was part of a vast desert, similar to the Sahara in climate and structure. As the previous Permian period drew to a close, large rivers spilled an assortment of sand, silt and gravel across the landscape, creating the English Midlands. Depositions of this alluvial material formed thick strata, which became the Keuper and Bunter sandstones and gravel beds.

The Common is sharply divided into a number of levels or terraces, the origin of which is open to conjecture. Upper Terrace is the highest, formed of drift deposits of coarse sand and gravel, and it has been suggested by Professor Wills, in a paper published in 1926, that it was deposited by glacial meltwater. Pockets of clay scattered among the coarser deposit support this theory.

Deep channels running down from the terrace are a characteristic feature of the Common and are known as periglacial valleys. Highly porous, the Triassic sandstone froze during the post-glacial period and created a permafrost which forced meltwater into channels, now left as dry valleys. An interesting feature of the slope-wash area of the Common is the presence of ventifacts. These are typically three-sided pebbles believe to be the product of anti-cyclonic winds, which arose as the glaciers retreated, sandblasting the pebbles with a shower of fine particles.

It would be easy to deduce from this description that the Common is a waterless landscape, but its proximity to the Severn and the presence of bands of clay ensure that some moisture has been retained. Evidence of this is best seen nowadays where the A4025 road crosses the Common. North of the road is a small bog with some open water, known as Rush Pool. This tiny area is Worcestershire's sole remaining acid bog and, since 1980, has been the focus of much attention because of the secrets it holds about the social and natural history of the heath.

In the early 1980s, John Tucker, working as an ecologist for the County Council, took a series of peat samples from the Rush Pool and analysed the pollen preserved within. Samples from these and a well-preserved pine log found in the depths were radio-carbon dated and estimated to be 10,000 years old. Pollen records showed that, when the pine fell, the common was wooded, predominantly with pine mixed with hazel. In 7700 BC, the wood may have harboured wolves, brown bears and beavers, while crossbills and crested tits bred in the pines, as they still do in lowland continental Europe. Man as yet had little influence on the landscape, which was afforested: it is said that a squirrel could travel from the Severn Estuary to the Wash without touching the ground.

By 630 BC, the pollen profiles showed a dramatic change. All trees were at relatively low levels: pine and elm had disappeared and alder and lime were in decline. For the first time grasses and agricultural weeds began to appear, evidence of human cultivation. Particularly interesting is the presence of buck's-horn plantain, then a weed of Iron Age crops, and now a classic pathside weed associated with the activities of man. The Common's landscape has since been shaped by human influence, though another animal has been equally responsible. In 1288, the Red Book of Worcester, a survey of the Bishop's manors, mentions a 'coney warren' in the area, the first evidence of rabbits. Poor land was often used for warrens, though these did not become established at Hartlebury until the mid-eighteenth century. Rabbits are now widespread, their grazing preserving many local species by keeping invasive vegetation at bay. Grazing by livestock was already causing problems in the sixteenth century. Leland, in his *Itinerary*, describes the area as 'enclosed ground woody and sandy but somewhat barren of corn'.

Stonechats visit heathlands regularly. (BW)

The Common's chief value to the manor of Hartlebury lay in its grazing rights and its timber, gravel and sand supplies. Extraction of minerals reached its heyday in the eighteenth and nineteenth centuries. Sandpits are still much in evidence on the Upper Terrace, but their wildlife interest is limited. In contrast, clay removed from the bog for brick-making created bare spots essential to the survival of plants such as round-leaved sundew and marsh club-moss. Rope-making was a local industry, using hemp as the principal source of fibre. The fibres were separated from the stem by 'retting' — soaking plants in pools. Cannabis plants have occurred in the vicinity in recent times, although their finders are naturally reticent about making their discoveries too public.

Hartlebury Common has long been a Mecca for naturalists. Edwin Lees, a celebrated Victorian botanist, published a graphic account of the Common's floral specialities. In 1856, his *Pictures of Nature* contained the following passage: 'This heath offers one of the pleasantest places for botanical rambling in Worcestershire, and is very fertile in Carices and bog plants. The great bog on the northern side of the heath presents a beautiful appearance in early summer, with the flossy tassels of cotton grass . . . white as driven snow; while later, when the pink bells of the *Erica tetralix* appear, and various ferns have unrolled their fronds, the scene is still more beautiful'.

In Lees' time the Common was botanically richer. Alpine, marsh and stagshorn clubmosses occurred together with all-seed, smoooth cats's-ear and white horehound. Botanists of the period became intrigued by the presence of a number of maritime plants, such as sea-stork's-bill and sea bindweed and attributed these records to old associations with the Severn estuary. By the advent of the early twentieth century these views were falling from favour.

Bird life on the Common was also well-documented, notably by the great ornithologist Eliot Howard, whose seminal work on bird behaviour *Territory in Bird Life* was published in 1920. His notebooks offer tantalising glimpses into contemporary bird life with records of wrynecks, cirl buntings, corncrakes and woodlarks, all lost to the county as breeding species.

Losses apart, Hartlebury Common is still a rich hunting ground. Now a Local Nature Reserve, it is managed by the County Council for its wildlife and recreation interest. Encroachment by trees is under control and Rush Pool has been partly cleared to encourage dragonflies. The open sandy areas have a specialist flora of early-flowering plants which flourish without competition. In April the ground in a few places is covered with a frosting of shepherd's cress, while common whitlow-grass has the same effect along paths. Here buck's-horn plantain, early forget-me-not and hoary cinquefoil also occur with heath groundsel later in the year. On each side of the A4025 there are strips of floriferous turf grazed by piebald ponies and rich in flowers. Yellow rattle, bird's-foot and spring vetch grow here, but the star attraction is sand catchfly. Numbers vary each year and this cerise-flowered annual has a tenuous hold on a narrow patch of verge. The plant occurs nowhere else in the Midlands and may have been unwittingly transported here by Romany travellers. Tucker and Betts in *Hartlebury Common; its social and natural history* note that a local Romany family, the Boswells, have relatives in the East Anglian stronghold of the catchfly.

Damper spots over the clay have a distinct flora which includes cross-leaved heath, purple moor grass and lousewort. These are all confined to the Lower Terrace where the bog is the most important area. Broad-leaved cotton grass is dominant in the wettest patches and a variety of sphagnum mosses cover the ground, including the local *Sphagnum squarrosum*. The cushions of moss contain Worcestershire's only colony of round-leaved sundew, an insectivore which may owe its presence to man's clay-digging activites; certainly plants are quick to colonise the bare depressions which still remain. Sadly, it is man who provides the greatest threat to these plants by removing the moss for use in hanging baskets. Marsh violets also grow sparingly in the wetter areas and the cyclamen-like blooms of cranberry can be seen in June. A plant to look for is marsh club-moss, whose yellowish stems lie prostrate on bare peat; although it has not been seen for many years there is always a chance that disturbance will activate dormant spores.

Unlike the bog, Rush Pool contains permanent water and so has a characteristic flora. Bog pondweed is rampant, belying its scarcity in the county, and marsh cinquefoil covers the surface with vinaceous flowers in summer. Among these are the delicate, fringed blooms of bog-bean, once collected and sold for threepence a bundle in Dudley as a cure for back pain. Around the pool margins grow the fleshy leaves of marsh pennywort.

Rush Pool is especially rich in invertebrates, which include the rare fenland water beetle *Hydraticus transversalis,* normally found in East Anglia — not the Boswells again, surely? Dragonflies are well represented; emperor and club-tailed have been recorded, and broad-bodied libellulas are frequent. Out on the heath the insects and spiders match those of the Devil's Spittleful, with a few additions. Two beetles are noticeable in late spring when they can be found stumbling along on short turf. Minotaur beetles are equipped with spectacular horns and lay their eggs in rabbit droppings. Bloody-nosed beetles are herbivores, feeding on bedstraw; when picked up they exude bright red droplets from their mouthparts.

The butterflies of the Common are not especially noteworthy, though moths present a different story. The list of species includes the beautiful yellow underwing, bordered gothic, emperor and small elephant hawk. Perhaps the most surprising moth is the Portland moth, a coastal species not seen since 1900 — another intriguing link wth the maritime theory.

Winter bird watching on the Common can be disappointing, though in the 1970s a sentinel great grey shrike could regularly be seen. Spring and summer are more rewarding and in a good season every gorse bush seems to contain a whitethroat, yellowhammer or linnet. In lowland Worcestershire meadow pipits are rare breeders: the Common holds up to five pairs, though broods are often lost to dogs walked through their main nesting areas. Stonechats are similarly affected. Their status here is that of a sporadic breeder: pairs bred last in 1977 and 1978. Whinchats and wheatears are now reduced to passage migrants, both benefiting from tree-clearance and regular fires which remove taller vegetation.

Fringing oakwoods have a typical bird population and the interface between wood and heath is the optimum habitat. Tree pipits and spotted flycatchers are characteristic species and sparrowhawks hunt along the wood margins.

Heathland has no special mammals, though rabbits are a prime influence on the landscape. Maybe some will be eaten by mink, which have become established at Hillditch in recent years.

Kingsford is the third significant area of heath in Worcestershire. Lying at the southern end of Kinver Edge, this country park has been largely planted with conifers so that a 30 acre remnant of heather and birch is all that persists. Its flora and fauna are more typical of open birch woodland. Bracken and wavy hair-grass dominate, interspersed with ling and bell-heather in drier areas. Along paths, the rush *Juncus sqarrosus* is a local plant in the county.

Kingsford's glory is its fungi. Birch has an impressive list of associated species and many can be seen here during an October foray. Most important of all is the birch polypore which attacks living trees, maintaining an open canopy woodland and providing nest-holes for redstarts and tits. The white-spotted red caps of fly agarics are magnificent at Kingsford, often growing in association with mustard-yellow earth balls. Other fungi include birch boletus, brown roll-rim, tawny grisette and charcoal burner, and many species must await

Birch bracket fungus is common on the old birches at the Devil's Spittleful. (LS)

discovery — scope for the keen mycologist.

In March day-flying orange underwing moths appear around the birch tops and summer days see antler moths skimming low over the heather. In autumn red admirals and commas are attracted to oozing birch sap. Almost every year someone is bitten by an adder at Kinver or Kingsford, even though the numbers of these snakes is low. They are best seen in spring when they emerge from hibernation and lie-out in undisturbed spots. Slow-worms are more common but difficult to detect among the undergrowth.

There are few, if any, exclusively heathland birds here. Until the mid-1980s nightjars strayed from the more suitable Staffordshire sites, although these have now vanished in line with national trends, despite careful habitat management. Green woodpeckers are much more common, with several pairs breeding in the birches and older oaks. Up to ten wood warblers hold territory and a May walk is punctuated by their song, a delicious shivering trill during which the bird vibrates with its efforts.

Worcestershire's remaining heathland sites are isolated fragments, constantly under threat from agricultural development. One such is Barnett Hill, the majority of which was destroyed to create horse paddocks. A tiny patch of sand and heather remains in which the rare wood cudweed lingers against all odds. Nearby, in the reclaimed fields, annual knawel is a reminder of a richer flora. A similar, but larger patch of heath lies a mile distant at Mount Segg. There are no rarities here, but common lizards are locally significant.

Another heathland lies in Habberley Valley, west of Franche, near Kidderminster. The area is bordered by sandstone 'cliffs' in the north. High on the bracken-covered slopes adders and slow-worms bask, and the creamy flowers of climbing corydalis twine among the undergrowth. Wall pennywort, wild service tree and yew sprout from the sandstone, though better plants are inaccessible. Habberley is popular with visitors and the more vulnerable species have been trampled out of existence: moonwort and wild pansy have probably gone forever. Two insects of note are the heath bumblebee, attracted by the abundant bilberries, and the purple-glossed click beetle *Ctenicera cuprea.*

We are fortunate in Worcestershire that our principal heathlands are so well-managed. Surrounding counties have corresponding fragments, though it may be too much to expect recolonisation.

Woodcock roding display flight over the Devil's Spittleful. (BW)

ABOVE and LEFT: The Devil's Spittleful rock, covered with old Scots pines, rises from the heather-covered heathland. (DMG) BELOW: Looking north from the Devil's Spittleful rock across the heath. (DMG)

ABOVE: The sand dunes of Hartlebury Common. (DMG) BELOW:
The Rushy Pool and bog at Hartlebury. (GHG)

The view from the Beacon on Lickey Hills shows the urban areas of Rubery, and Bartley Reservoir on the northern county boundary. (DMG)

Wildlife in Town and Village

A county whose symbols are apples, pears and Worcestershire sauce is not the first which springs to mind when one thinks of urban wildlife. However, the Industrial Revolution began with a vengeance in the Old County, which in the late eighteenth century included Dudley and large chunks of the modern 'Black Country'. The fumes from a thousand chimneys and furnaces filled the sky and it is said that Queen Victoria, travelling through the area by train, had the blinds of her carriage pulled down lest she be offended by the industrial chaos outside. Writing later, Pevsner extolled the virtues of the rural county, likening it to a pie, of which Dudley was a piece of burnt crust.

In 1975, the creation of the new county boundaries excised Worcestershire from the West Midlands conurbation and considerably reduced the county's urban acreage. The principlal towns are now Worcester, Kidderminster, Bromsgrove, Droitwich, Evesham, Pershore, Malvern, Stourport and Redditch, while Stourbridge has an honourable claim for inclusion as it lies within the Watsonian vice-county (used to record wildlife) and is almost unanimously regarded by its residents as part of the Old County.

Modern Worcestershire towns comprise an historic core surrounded by rings of housing and large industrial estates which have mushroomed in the last twenty years. Within built-up areas are a variety of habitats for wildlife, some deliberately created, but the vast majority accidental. Of all Worcestershire towns Redditch has experienced the greatest change in size, being designated a New Town in the 1960s. Despite its great increase care has been taken to maintain open space within the New Town along the Arrow Valley and at Bordesley. These areas contain interesting wetlands, including the Arrow Valley Lake and old fish ponds. Remnants of old pastureland occur within the town as well as the Ipsley Alders Marsh.

Worcester itself demonstrates the richness ofo older towns for natural history and this was graphically revealed by a survey of the city's green spaces in 1988. The WNCT and the

Cormorants are common visitors to the Severn in Worcester, 'hanging our' wings to dry on the lock gates, cathedral tower or on the tall spire of St Andrew's 'Glover's needle'. (RMB)

Community Landscape Trust, aided by funds from the government's Manpower Services Commission, prepared a detailed report showing the city to be richer in wildlife than many parts of the surrounding countryside.

Worcester City boundary encompasses 3,334 ha and the combined area of all green spaces over 0.5 ha is 1,667 ha or about half the total. If smaller sites and small gardens are included, up to 70% of Worcester is green! This astonishingly high figure makes Worcester one of the greenest cities in Britain and, although this area will shrink with the building of new large estates at Wardon, Worcester will remain well and truly green. The team of surveyors were constantly surprised by the wealth of botanical finds. 515 species of vascular plants, 85% of the county's total were recorded, largely due to the diversity of habitats. The records included such rarities as marsh stitchwort, dropwort, spiny restharrow and corn buttercup. Twenty species of butterflies were found, including a colony of white-letter hairstreaks and white admirals in Nunnery Wood. Birds were well represented, with over 70 species, from the expected house sparrows to a storm-blown marine manx shearwater. The city was revealed as a nationally important site for great crested newts, found in fourteen ponds. Add to this six badger setts and four species of bats and the result is an excellent wildlife resource on the doorsteps of many thousands of fortunate residents.

The Severn provides Worcester with a natural corridor along which wildlife can pass into the heart of the city. The river is prone to flooding and this has curtailed development on lower land. Alders fringe the banks below the Cathedral, where a motley collection of plants includes three species of figwort, with probably the most accessible green figwort in the county. Adventives, the alien plants which grow from imported seed, are common. Hops and flax may derive from old warehousing, but canary grass originates from birdseed thrown by passers-by. Siskins feed in the alders, showering winter walkers with a fine dust of seeds near the river bridge. Pitchcroft Racecourse is attractive to gulls during floods and they can be watched probing water margins for drowned invertebrates. In summer the planted limes shelter an unexpected colony of broad-leaved helleborines. Birds continue to use the river as a migratory route regardless of urban development. Occasionally the Cathedral tower is adopted as a resting post for a wing-drying cormorant, whose heraldic pose suggests that it was part of the stonemason's original design. Terns have also been noted and swallows, not usually urban birds, skim over the water.

Unimproved grassland in the city will soon be reduced from its present 48 ha with the construction of the Wardon estate, but a rich block will remain near Nunnery Wood. This has recently come under the stewardship of the County Council's Countryside Services and will be managed for its botanical and entomological interest. At Elbury Mount and Rainbow Hill, meadow is rapidly grading into scrub, threatening traditional grassland plants. Taller and more robust species survive here, including the narrow-leaved everlasting pea, wild liquorice and the occasionaly woolly thistle, all highly respectable urban plants.

There are 49 ha of broad-leaved woodlands in the city, mostly on ancient woodland sites at Perry, Nunnery, Warndon and Tolladine Woods. At the last locality there is a stark contrast between the modern housing estates and the surviving block of trees. Wild service testifies to the antiquity of the hedge and the wood itself harbours a botanical gem, the green hellebore whose improbable emerald flowers appear in March when few other plants bloom. Nunnery and Perry Woods, now part of Worcester Woods Country Park, are pedunculate oak with a rich flora. Perry Wood has a fine colony of regenerating hornbeams which are almost certainly planted, but attract hawfinches nevertheless. Wild cherries, abundant field maple and a few small-leaved limes make this an exceptional city woodland. It is bordered by estates to the north, and residents make frequent sorties into the wood to gather illicit timber. Though indiscriminate cutting can ruin localised areas, there is a certain aptness in the notion of twentieth century city dwellers gathering firewood as their ancestors did in medieval times, possibly from the original stock. Community woodlands are undergoing something of a resurgence at present, and there is undoubtedly scope for their inclusion in urban planning schemes. Nunnery Wood is owned by Christchurch College, Oxford and the Regional Health Authority and is a valuable recreation area. Interestingly the woodland floor is ridge-and-furrowed and wild service is almost confined to the perimeter, suggesting a history of cultivation sometime in the past.

Marsh is a rare habitat in the city, although an area at Northwick is worthy of designation as an SSSI. There are extensive areas of wetland habitats at Diglis and a series of medieval fishponds at Lower Wick. 75% of the ponds in Worcester city contain amphibians, compared with only 35% on farmland.

A walk through Worcester City can be remarkably rewarding, producing more records than in intensely farmed areas. While this was being written a lady phoned to ask 'would we like a beautiful dead crow!' We could think of no immediate use for one, but it transpired that she lived in the middle of Worcester and the crow nested in a huge plane tree outside her window. The garden below received gifts of droppings, bits of dead fish and other leftovers from the carrion crow's table above! She had seen thirty-two bird species in her small garden and in winter was often rewarded by flocks of fieldfares and redwings descending for berries. Opportunistic mice had built a nest near the bird table to take advantage of a food resource which arrived from above, like manna from heaven. This small example illustrates the value of urban green space for wildlife, and of the value of wildlife to the people who live in town.

Worcester has a great opportunity to use its green spaces for conservation, education and public amenity. Through correct management, these sites could become an environmental resource for the city's thirty-five educational establishments and 25,000 children, giving them a chance to get involved with natural history on their doorsteps. The same can be said of all towns and villages in the county.

Urban habitats vary considerably in their attractiveness to wildlife. New housing estates offer limited opportunities for plants and animals to gain a foothold and it is the hardiest or most resourceful which win through. Old parks and cemeteries are often havens for wildlife since they are relatively undisturbed and have been in existence for many years.

You don't need to travel far to find wildlife — a great deal shares the same building with you! Man-made structures support a surprising number of plants, the list increasing with the age and depending on the composition of the structure. Worcester's oldest walls, in Cathedral Close, are topped by wild wallflowers, introduced centuries ago from mainland Europe. Mediterranean snapdragons also flourish in the city, their maroon spires burgeoning from cracks in the walls of older buildings, producing an unexpected midsummer display. Another alien has, literally, found a niche. Buddleia, the butterfly bush, sprouts everywhere

Buddleia is the 'butterfly bush' of derelict industrial sites, here with red admiral. (TP)

on disused lots, often ascending high up walls and luring butterflies into partly demolished houses. Small tortoiseshells, peacocks and commas are no strangers to our dwellings and hibernate in the cooler areas of our homes, emerging in warm spells to flutter at windows. Typical wall-loving plants include ivy-leaved toadflax, pellitory-of-the-wall and a number of ferns. Leaking downpipes provide a site for spore germination and shady town-centre alleys can field a rich supply of ferns. Male ferns, broad-buckler and hart's-tongue ferns are common in such places, often growing with delicate fronds of bracken which looks completely different from its normal robust appearance. Sunnier spots support wall-rue spleenwort, a frost and drought-resistant species, especially fond of wall-sides where lime mortar has perished, and occasionally maidenhair and black spleenworts. One of the prettiest ferns is the rusty-back, a rare Worcestershire plant occurring with regularity in the Stourbridge area on the blue brick walls of the Black Country. To those botanists familiar with this plant on western walls, it can be a surprise to find it skulking in a back street.

Spiders take full advantage of man-made structures as many arachnophobes will confirm. The large house-(or should they be called bath-) spiders are to some extent dependent on man and readily take up residence in our homes, entering through cracks and not *via* the plug-hole. A particularly leggy spider is *Psilochorus simoni,* not yet recorded in the county, but the late Dr Douglas Cowden was convinced it lurked in wine cellars beneath Worcester city among the bottles of Mouton-Cadet. Discovery of *Psilochorus* offers a research opportunity for the steadfast arachnologist.

Bats have almost as few fans as spiders though, happily, their followers are growing in number. Worcestershire has an active Bat Group, whose members conduct surveys and advise householders on how to preserve their co-habitees, now fully protected by law. A sign of changing attitude was a recent advertisement by a Malvern estate agent of a property with a colony of lesser horseshoe bats, obviously viewed in the same light as a fitted bedroom or loft conversion. Better by far to have bats than a loft conversion. Contrary to popular belief, bats like clean, dry resting places and have taken to modern housing well, often roosting behind slates or weatherboarding. Pipistrelles are still the commonest urban bat, though in the Kidderminster area, Daubenton's bats skim low over nearly every watercourse and noctules roost in factory chimneys. When a Kidderminster chimney was demolished in the late 1980s members of the Worcestershire Bat Group saved over eighty bats from a roost of about 100 — the remainder flew off safely. Other urban bats include brown long-eared bats roosting in a car showroom in Kidderminster and a rare barbastelle found dead at a church roost.

Birds use buildings as nest sites and many frustrated property owners have been forced to summon the sweep to remove a jackdaw's tangle of twigs from their flues. In the Worcester city centre, lesser black-backed gulls breed on the rooftops, the adults scavenging at local tips, which remove their reliance on sea-food. Black redstarts have also adapted to

Goldfinches feed on thistle seed wherever it grows in town. (WNCT)

townscapes, where they find a replacement habitat for mountain crags. In Worcestershire they remain rare; a pair summered in Worcester in 1961 and breeding was attempted in a canalside warehouse in Stourbridge eleven years ago. A male sang from an Evesham factory roof in the 1970s. An equally elegant city dweller is the grey wagtail which occurs in built-up areas in autumn and winter.

The importance of gardens as wildlife habitats can hardly be over-emphasised. Stand on a vantage point such as Bredon Hill and you will soon appreciate why. Gardens stand out as enclaves of green amid a landscape of huge fields and scattered hedges. The larger, unmanaged gardens of Victorian villas offer the best sanctuary for wildlife, though these are now being sub-divided as the pressure on building land increases. Two groups have become reliant on gardens – amphibians and birds. The common frog no longer lives up to its epithet in many parts of the county, but is undoubtedly more frequent in garden ponds than in the surrounding countryside. Newts too, prosper in ponds, reaching high densities if goldfish will allow. In Worcester, slow worms are a feature of many garden compost heaps, where they perform a valuable service eating slugs and small snails.

Woodland birds have adapted well to gardens and many familiar species have their origins in ancient forests; the archetypal garden bird, the robin, is a shy woodland species in Europe. Blackbirds, song thrushes, tits, nuthatches and tree creepers are in this category, as well as most finches and crows. A notable exception is the collared dove, whose spectacular change in range in the early twentieth century has yet to be satisfactorily explained. Originally from Asia, the doves spread phenomenally in the 1930s, spreading north-westwards through Europe and reaching Britain in 1955. Worcestershire's first record was in 1961 when a pair bred near Worcester. The doves breed throughout the year and only three years later flocks of fifty birds were reported. Numbers have now stabilised and many gardens have breeding pairs, especially where conifers used as nest sites are well-established.

Feeding birds is part of the British way of life and contributes significantly to the survival of many species. Blue and great tits are permanently associated with peanut dispensers, appearing from nowhere when a fresh supply of food is put out. In recent winters red bags of nuts have been enlived by siskins, which turn to this food supply when alder and birch seeds are exhausted. Feed the birds and you unwittingly feed the predators. Many a shocked householder has witnessed the demise of a garden favourite, borne away by a visiting sparrowhawk. Once scarce in the county, this superb raptor has increased following the decline in the use of organo-chlorine pesticides, and it is now very much an urban bird. Its hunting strategy relies on surprise and speed, leading some unfortunates into windows.

Gardeners generally welcome birds and usually, frogs and newts, but are less kind to the creatures on which these animals feed. War is mercilessly waged on a variety of garden pests, even though there is no shortage of natural predators. Weeds suffer but by their very nature are equipped to fight back. Some like hairy bittercress reproduce in huge numbers; others like field horsetail have deep roots and are well-nigh impossible to eradicate. A few are even desirable. Caper spurge is a rare native of cleared coppice in England and occurs as a garden weed in the county. This vegetable pagoda reaches five feet in height, flowering in its second year. Its capers are a powerful diuretic and the stem bleeds an astringent milk, pumping out the fluid as though from an artery. Worcester gardens are characterised by annual mercury, a cousin of the woodland perennial dog's mercury.

During a survey by WNCT in 1977-78, 88 ha of the county, comprising 200 sites, were covered by churchyards. A rich flora and fauna may be associated with churchyards as many are relatively undisturbed and cared for by environmentally sound methods. But there has been a worrying series of best-kept churchyard competitions of late which do not auger well for the prospects of ecclesiastical wildlife in the future. Trees around churches are valuable nest sites for many birds, including goldcrest and coal tits, which are associated with conifers, and thrushes and woodpeckers. Spotted flycatchers are characteristic graveyard birds and use headstones for perches, sallying forth to seize flying insects with an audible bill-snap. Gravestones are important substrates for lichens, though these are easily killed by industrial pollution. Stone walls are often the best habitat for miles around: one Worcestershire churchyard wall carries nine species of fern, numerous mosses, stonecrop and wall pennywort. Malvern Priory has good colonies of the last, growing with maidenhair spleenwort and shining crane's-bill. There are strong similarities between old meadows and churchyard grasslands and, near Droitwich, ox-eye daisy, rough hawkbit, cowslip and hoary plaintain demonstrate the neutral nature of the soil and the previous flora of lost pastures.

The church is a sanctuary for many birds. Jackdaws, starlings and swifts nest in bell-towers, along with kestrels. Worcester Cathedral is often enlivened by this falcon and busy flights of jackdaws. Screaming parties of swifts are common across towns in the county, where these 'avian mobsters', as Ted Hughes called them, are entirely dependent on buildings for nest sites, using holes made by missing tiles or in stonework. House martins build their mud-pellet nests beneath eaves of buildings in towns and villages, occasionally finding a more substantial building material; such a bird used wet concrete and built a nest so strong 'that even a boy with a tennis ball could not dislodge it at the end of the summer'. That impregnable fortress must have prevented the usual take-over bids from sparrows.

Factories and trading estates may appear at first glance to be unexciting but can be rewarding to the keen botanist. At Droitwich, prickly lettuce and field madder occur, while at Blackpole, in Worcester, corn buttercup sprouted on newly disturbed ground alongside grass vetchling. Sandy Kidderminster soils produce a different crop of ruderals, including hare's-foot clover, kidney vetch and tree lupin. Modern warehouses are used as pied wagtail roosts and in 1985 a tawny owl entered such a roost to pick off sleepy wagtails.

Waste ground is similar in many parts of the English Midlands. Inner city areas are rapidly colonised by plants with wind-blown seeds. Composites are frequent and include the classic alien, Oxford ragwort. This plant is probably a native of volcanic rubble in Sicily and escaped from the Oxford Botanic Gardens to be wafted round the country in the slipstreams of trains and, eventually, cars. In 1909 it was locally abundant, occurring at Shrub Hill railway station in Worcester and at Pershore and Droitwich. Nowadays it is still an urban plant, especially associated with the Black Country.

Another county success is the rosebay willow herb. Amphlett & Rea in 1909 noted that it was 'a handsome plant frequently grown in gardens, but not so frequently seen in the wild'. How things have changed! Few people give garden space to this invasive but attractive flower now. Wormwood is another common wasteground plant, forming greyish tufts over large areas. Its acrid leaves were chewed by Black Countrymen on hot days to induce salivation and slake their thirst. The spread of these three plants has helped their attendant moths. Wormwood sharks have increased in towns and the elephant hawk moth is occasionally seen hovering round clumps of rosebay. The hawk moth's huge caterpillars are snake mimics

Rubbish tips attract thousands of gulls to feed on our thrown-away take-aways. (RMB)

and a not infrequent source of enquiries to WNCT office. Oxford ragwort is the food plant of the cinnabar moth, a brilliant carmine and black daytime flier. The adult mimics the poisonous burnet moths and its tiger-striped larvae display the warning black and orange stripes of a wasp.

Town and city sports fields sometimes turn up trumps, though it would be hard to beat the rough-legged buzzard which perched on rugby posts at Bromsgrove School in November 1974. Wheatears on passage and wintering thrushes are regular, appreciating the churned-up mud of goal-mouths. Gulls roost on football pitches, especially if these are close to rubbish tips.

Rubbish tips are rarely placed in or near towns nowadays, although it is only a few years since the large tip in Worcester, near Hallow, was filled and closed. Tips attract large numbers of birds, especially in winter, and huge flocks of black-headed and lesser black-backed gulls are common at Worcestershire's tips, joined by smaller numbers of herring, great black-backed, and common gulls and occasional glaucous and Icelandic gulls. Throckmorton tip produced the county's first ring-billed gull, a North American species, and Madeley tip has attracted Iceland and Mediterranean gulls. Bird-ringing studies have shown that winter black-headed gulls are from European, from as far away as Russia and the Baltic States. Most of the lesser black-backs breed at colonies in northern England, though a few come from Iceland. Tips also attract large numbers of scavenging starlings, carrion crows, rooks, jackdaws, foxes and rats and the warmth of rotting organic material attracts house crickets.

Filled tips are now usually capped with soil and nicely seeded with grass but, in the past and if this is not done, large expanses of weeds grow and their seed crop is sought by flocks of finches throughout the year. The broken terrain at tips is attractive to wintering stonechats and the more open areas to pied and grey wagtails, skylarks and meadow pipits. All this activity brings in predatory weasels, stoats, kestrels and sparrowhawks.

In the 1980s, work in Worcestershire helped to show that gulls could carry bacterial diseases caused by *Salmonellae* and *Campylobacter*. Indeed *Campylobacter laridis* was so named after *Larus,* the generic name for gulls, following descriptive work at Worcester Public Health Laboratory.

ABOVE LEFT: Quail, an occasional summer visitor. RIGHT: On the way out — corn cockle, and CENTRE: on the way in — green nightshade, a South American newcomer. BELOW LEFT: Rape in full bloom, an increasingly common sight. RIGHT: Corn flower and field poppy. (All DH)

LEFT: Grassy churchyard and lichen-encrusted gravestones. (DMG) RIGHT: The Worcester Canal between factories — a wildlife pathway in the city. (DMG) BELOW: Worcester is visited by great floods which the gulls enjoy. The viaduct and its lower partner provide niches for many plants in the city. (DMG)

Hedge-laying at Upton Warren reserve. (DMG)

The Trust

The Worcestershire Nature Conservation Trust is twenty-two years old in 1990. It was founded by a group of dedicated naturalists and planners, led by Christopher Cadbury, when the West Midlands Conservation Trust re-formed into separate county Trust in 1968. Since then membership has grown to over 6,000 and the Trust has established nearly 60 nature reserves, covering 1,500 acres. In 1990 it moved to a new headquarters at Lower Smite Farm, near Worcester, where at long last it has enough room for its offices, records and future developments.

During its first twenty-two years the Trust has built up files on the natural history of the county, believing that it can pursue its aims most effectively from a sound base of knowledge of the county and its wildlife. In the face of massive agricultural change the Trust has established many reserves covering some of the best sites in the county, aiming to conserve good examples of the rich variety of Worcestershire's countryside. Its early policy was to acquire SSSIs as these were the best habitats and in need of greatest protection. Money could be raised from grants and donations to buy SSSIs whereas funds were not easily available for unscheduled sites. Nevertheless acquisitions of SSSIs was and is a large task as Worcestershire contains more than most other counties, a reflection of the diversity and importance of its habitats. SSSIs are perhaps better protected by Wildlife & Countryside law these days, although recent studies in Worcestershire and throughout Britain show a continuing story of damage to SSSIs, the country's most important wildlife sites.

The reserves lie at the heart of the Trust's affairs, not merely as protected sites, but because they are natural history showcases where people can go to see the wonders of the county's wildlife, and to enjoy them. Such experiences must, surely, stimulate a wish to help conserve these and similar places. Most people, especially children, have an intrinsic interest in the natural world. This is apparent on reserve Open Days when guided walks by naturalists are one of the most popular features. The reserves also demonstrate management methods to other owners and show that in suitable circumstances it is possible for economic forestry and farming to go hand-in-hand with conservation. There are other ways of managing woods than converting them to conifer plantations or following the edicts of custom. Perhaps a visit to a woodland reserve will stimulate entrepreneurial flair and lead the way to economic coppice regimes? WNCT is leading the way in applying silvicultural science to broad-leaved woodland management with conservation in mind. The Trust wishes to conserve ancient woodlands but the money to do so has to be raised and forestry regimes which produce good timber and are good for nature conservation have to be found. It is not feasible to completely return to large acreages of coppice woodland: some must be established but other approaches need to be investigated.

Science must play a part in nature conservation. The fire of emotion is fine but the fire must be fuelled by factual information whenever possible. Understanding of the ecology of plant and animal communities is essential for their proper conservation, but obtaining the understanding involves many years of painstaking research. The pace of change is so great nowadays that one cannot often wait for the science; pragmatic decisions have to be taken, based on experience and knowledge. Repeating old management regimes will, perhaps, produce the same results as in the past, whether in woodland or meadow.

The Trust has long worried about conservation in the wider countryside, and the vulnerability of many sites of great interest but below the SSSI grading. These form a greater mass of countryside than SSSIs and they may either be conserved or destroyed without the owner's realisation of the importance of a site. This has led the Trust to build up a register of Special Wildlife Sites which should help conserve this second tier.

'Partnership' is a popular word in county conservation Trusts at present. All the county wildlife trust and urban wildlife groups are corporate members of the the Royal Society for Nature Conservation, the national umbrella organisation. The RSNC and the Trusts are now described as 'RSNC, the wildlife Trusts partnership'. This reflects the dual role of all concerned: wildlife conservation at the local, county or urban level, coupled with an awareness of national problems and the need for national action. The Partnership is steadily increasing its influence in matters of wildlife conservation in the United Kingdom and has the potential for even greater effectiveness in the future. It is based on charity and voluntary effort. The Trusts are all registered charities and work for the good of the cause without personal gain for the members.

The members of the Worcerstershire Trust have voluntarily provided most of the workforce during the last twenty-two years, working with a small paid staff of two or three which is all that could be afforded. This successful partnership between staff and volunteers has achieved much and will continue to be the bedrock on which the Trust is based. However, its affairs are becoming increasingly complex and demanding, beyond the time-capacity of a volunteer, who must work elsewhere for his or her living, and more staff will be required for the Trust to continue developing. For a few years many people worked for the Trust under the government's Manpower Service Commission schemes for those who could not find work. This valuable contribution to the Trust's work is now finished but for some years it helped progress in many ways and, hopefully, set many people on the road of working for nature conservation.

The challenges facing the Trust are great. It is often stated that one aim should be that conservation of nature takes a high priority on at least 10% of the county's land. It is unlikely that the Trust could, or would want to manage an estate of 17,500 ha. But it should work with others to encourage progress towards this ideal. What is needed is an interconnecting web of wildlife running throughout the county with particular jewels here and there — the nature reserves. The conserved areas would of course be in either private or public ownership and would be for people to enjoy as well as for nature conservation. An important characterstic of the county Trust conservation movement is to achieve nature conservation by working with people and by example. Although particular campaigns are useful to raise awareness, the Trusts are not primarily campaigning organisations, in the sense that they do not often put on demonstrations and wave banners. Nevertheless, the Trust cannot succeed by compromising. Half a site lost is gone forever and wildlife and nature are always the losers.

So much has been lost. Can we regain lost ground? Worcestershire has been ravaged, though it still contains much of wildlife value. To achieve the 10% and more the conservation movement has to catch the hearts and minds of all those who look after the land, so that conservation is second nature rather than an afterthought. To succeed in this objective requires persistence and education in the widest sense, through publicity, in schools, and with younger children through the WATCH club.

Sadly, money counts and, even with cash, progress will be slow; without cash it will be slower still. The Trust has to continue to raise money. In the Trust's history there have been several consistent and generous benefactors and without them progress would have been minimal. Many volunteers have worked hard to raise cash through all the avenues charities use to collect money, from shaking tins on street corners to writing books. Without volunteers, progress would have been even slower. In the New Age all these things will have to continue together with a much cannier input of marketing skills to enable the Trust to tap new sources of support from sponsorship to corporate membership.

Another challenge lies in towns. Urban wildlife conservation and the use of 'green space' is an important issue. Recent studies have revealed more wildlife in town than in much of the countryside; rare species lurk on municipal car-parks and on warehouse roofs. Schools in town cannot afford to send children to the countryside for natural history or biology lessons, yet we now realise the resource lies at hand, ready for use, not a stone's throw from the school's doorstep. Some of us had the good fortune to be taught by biologists who could see lessons on animal and plant behaviour, ecology and evolution on a stone wall or in a horse trough (not many of those about nowadays!), and could reveal this magic to children and capture their imagination with an interest leading to a lifetime's enjoyment. Think what could be done in Worcester, where there is 50% open green space and 25,000 children — all potential nature conservationists. Also, of course, practical conservation starts on your doorstep. In the country you may be a farmer with 600 acres to play with, but in the town it is more likely to be your garden, or the park down the road, or the canal bank, or the scruffy patch of land left by builders, or the churchyard. The Trust wants to encourage everyone to open their eyes to the natural history of these sites and to encourage them to start nature conservation at home.

To succeed and develop further, WNCT needs many friends and supporters. It is likely that you, gentle reader, already support the Trust. If you do, please accept our thanks for your help and for buying this book, as the royalties go to the Trust. If you are a member of the Trust please stay with us; if you are not, please join and help conserve a remarkable county before it is too late.

Children learn the secrets of freshwater life at an open day at the Knapp & Papermill reserve. (DMG)

Index

This index is mainly of species, places, people and a few topics. Broad references to topics can be found in the contents list. Figures in *italics* refer to illustrations.

Subscribers

Presentation Copies

1 Worcestershire Nature Conservation Trust
2 Royal Society for Nature Conservation
3 Nature Conservancy Council
4 Worcester & Hereford County Council
5 Worcester City Library
6 Christopher Cadbury CBE

7 G.H. Green
8 Brian Adlam
9 Brett Westwood
10 R.N. Bishop
11 D.N. Green
12 D. Hollis
13 Carolyn & Clive Birch
14 David N. Robinson
15 Brian Bentley
16 D.J. Cannings-Bushell
17 Michael David Southall
18 Michael F. Oldham
19 Digby Wood
20 N.L. & S.J. Boneham
21 C.S.R. Hartley-Cocks
22 J.H. Myring
23 K.J. Rogers
24 Brian Yates
25 Anthony Plowright
26 A.W. Brand
27 Andrew Clark
28
29 R.M. Bishop
30 R.T. McAndrew
31 P.D. Dickin
32 Stephen J. Green
33 B.L. Newman
34
35 Mr & Mrs D. Jennings
36 Lawrence Bee
37 Prof Ronald Gulliford
38 Michael Deighan
39 John Handy
40 Bernard & Olive Poultney
41 Gordon & Corrine Bailey
42 Dr Marshall Wilson
43 Charles Holmes
44 R.L. Averis
45 Gillian Coupe
46 R.S. Woodroofe
47 B.J. Gealer
48 Joan M. Wakeman
49 R.O. Ibrahim
50 Neil Waddingham
51 Stephen J. Gledhill
52 Jay Field
53 Canon G. Neville
54 Miss B. Swan
55 John M. Price
56 Richard Pitman
57 N.F. Tapscott
58 C.J. Wolstencroft
59 A.J. Tame
60 N.R. Bennett
61 Ms E.H. Morgan
62 M. Smallwood
63 Mr & Mrs B.K. Doling
64 R.A. Ogden
65 Cyril Kershaw
66 Mrs Valerie Harber
67 E.B. Brown
68 Tony Dorricott
69 George C. Smith
70 Arthur Royle
71 D.R. Bayliss
72 Martin J.N. McDonagh
73 David Carleton
74 Christine Loader
75 Clare Bishop
76 Graham Wynne Morgan
77 G.D. Clemmow
78 James Wood
79 Leslie Charles Pitt
80 Patricia Mary Keighley
81 Phyllis M. Walford
82 D.C. Lindsay
83 Michael Bryan
84 Neil Edwin Sambrooks
85 Mrs Judith R. Vaughan
86 Peter K. Dedicoat
87 Mrs J.M. Blackett
88 Air Commodore Peter Hughes
89 Miss B. Chatwin
90 Michael T. Averill
91 Shelagh & Geoffrey Peters
92 H. Mason
93 Brian Iles
94 Mrs Marjorie E. Clark
95 H.F. Hopkins
96 D.C. Kearney
97 Michael Jeeves
98 Y.E. Green
99 Joan Lever
100 Michael H. Bourne
101 N.N. Beach
102 Mark A. Rowe
103 Mrs C.M. Lewis
104 A.E. Gray
105 Jean Glazebrook
106 Dr G.P.G. Rowe
107 John Sirrett
108 Leslie H. Pinkess
109 J.C. Waine
110 J.W. Meiklejohn
111 David & Susan Matthews
112 Robin & Sarah Whittaker
113 Gordon H. Trengrouse
114 J.T. Hammond
115 A.F. Jacobs
116 G.L. Hollingworth
117 B.L. Drewitt
118 Marianne J. Wade
119 Keith Barnett
120 A.C. Wakeman
121 Gordon Forrest
122 Mrs P.D. Rogers
123 B. & M.D.H. Wright
124 John Dean
125 Arthur Pollard
126 W. Vanner
127 Brian J. Garrett
128 Mrs E.M. Whitlock
129 Henry W. King
130 Mrs Joan L. Burton
131 R.L.P. Milburn
132 Mrs J.T. Hart
133 G.D. Steer
134
135 Margaret Shipp
136 Derek E. Davis
137 Dorothy Bayliss
138 Mr & Mrs M. Chesterman
139 Rosemary I. Keep
140 George A. Wilkes
141 Elizabeth L. Kaye
142 M.A. & K. Parnell
143 Elizabeth A. Northover
144 M.J.A. Marsh
145 David N. Colclough
146 Mrs D.M. Jinks
147 Laure Read
148 Peter Whittaker
149 John Richard Copper
150 Dr C.M. Elliott
151 B.J. Wright
152 Jane Wise
153 H.J. Instom
154 Robert Everiss
155 F. Fincher MBE
156 John Rhymer
157 Beryl Haskins
158 Mrs Dicky Nedwell
159 Robert Deane
160 Jane Ratcliffe
161 Mrs M.O. Aspland
162 Neil Schofield
163 J.G. Hemming
164 J.M. Taylor
165 Mr & Mrs D.R. Hill
166 Mrs E.M. Humpage
167 Dr W.J.A. Payne
168 Mary Daulman
169 P.G. Garner
170 W. Kingston
171 Roger Robinson
172 Trevor G.C. Duke BSc
173 R. Gulliford
174 Mrs Ursula Gleenie
175 R.A. Vizor
176 Jan Harrison
177 Joy Pygott
178 Michael Savery
179 T. Evans
180 Andrew D. Shakespeare
181 Roger A. Summerfield
182 James Newton
183 H.D. Clayton
184 Mike Williams
185 Brian Bentley
186 John L. Oakley
187 Claire E. Turner
188 Dr F.M. Slater
189 Mrs H. Pogsmore
190 Mrs M.A. Caldwell
191 A. Bunney
192 Frank Baylis
193 D. Atkinson
194 R.A. Banks
195 S.M. Haynes
196 J. & M.R. Meech
197 H.E.E. Babb
198 Mrs J.F. Lambourne
199 S. Collings
200 M.R. Young
201 Desmond A. Jennings
202 James N.R. Blair
203 John M. Bailey
204 Vivien Allen
205 Dina & Ted Cockerill
206 Olive & John Thoumine
207 Werner Keis
208 D.R. Smith
209 Kenneth Bolter
210 R.J. Brant
211 Mrs M.L. Bentley-Taylor
212 Andrew R. Ward
213 Mrs Dorothy M. Green
214 Kay Bellbridge
215 Peter T. Humby
216 Mrs Rosemary Burton
217 Lindsay E. Drew
218 Timothy David Hugh Bashford
219 John & Jean Phillips
220 Joe & Ann Hardman
221 Dora Mottram Hayward
222 J.R. Hodson
223 Derek Brown
224 H. Cooper
225 Alf L. Dimmock
226 Dr D.J. Gobbett
227 Linda Sealey
228 R.W.D. Pannell
229 Mrs E. Tangye
230 A.E. Morris
231 R.A. Felton
232 Miss N.M. Whitehead
233 M.J. Billings

234 Eleanor Lloyd
235 Edward Probert Evans
236 Ann Gowers
237 Pershore College of Horticulture
238 Richard Lane
239 Anthony James
240 Lucy Harris
241 Peter J. Hancocks
242 D.F. Knight
243 Mr & Mrs D. Thomas
244 John Boyes
245 John Meredith
246 Mrs S.J. Wilkinson
247 Richard Coen
248 John R. Cockbill
249 F.M. Airley
250 Colin Grinnell
251 Barbara C. Redwood
252 Jennifer Turner
253 Brian H. Cackett
254 Kevin Andrew Heede
255 Dr Michael Hemming
256 R.J. Waine
257 Rachael Webb
258 J.H. Myring
259 Mrs J.E. Thorpe
260 Mrs A. Folwell
261 F.E. Skinner
262 Cathryn Pritchard
263 Martin Hales
264 Mrs Barbara Taylor
265 Alan F. Rankin
266 Bishop & Mrs Robin Woods
267 Mrs M. Tilt
268 Ian Nex
269 Heather Linley
270 S. John Savage
271 David Kingston
272 Raymond J. Hume
273 Gillian Harvey
274 Mrs Margaret E. Rodway
275 Doreen Margaret Cooke
276 David M. Ballard
277 Michael Lock
278 E.B. Watson
279 T.J. Pitts-Tucker
280 The Baskeyfields
281 Marion Baines
282 R.S. Bannister
283 T.F.C. Crawley
284 T. Baron
285 Tessa Carrick
286 Yvonne M. Way
287 Carolyn Partridge
288 Mrs W.M. Malvern
289 V. Mary Walker
290 Paul James Stanley Dunn
291 Peter Thomson
292 William Robert Leadbetter
293 Barry R. Freeman
294 Tim & Lesley Cox
295 Guy & Eleanor Busher
296 Jacquelynn Owen
297 Briony Serrell
298 Eddie Amess
299 Edward Walker
300 Anthony Wharton
301 Robert Drewrey Holdsworth
302 D.V. Gower MBE
303 Dr & Mrs E.M. Allen
304 David Philip Tabberner
305 Dulcie K. Salter
306 Mrs D. Simkins
307 Anne P. Daly
308 T.J. Higgins
309 W.D. Francis
310 Pat Orwin
311 Lionel Guest
312 Brian & Jacky Burrows
313 John Wridgway
314 David Robinson
315 Kelvin Hardiman
316 Paul & Julia Clarke
317 Mr & Mrs M.L. Burgess
318 N.J. Benbow
319 Richard Peplow
320 M.K. Rist
321 Michael Poulton
322 Elizabeth Mary Brown
323 S. & M. Leadley
324 Mr & Mrs A. Heaton
325 Rosalind Davies
326 Miss K.S. Lowe
327 Dr G.R. Martin
328 H.D. & W.A. Gardner
329 Mrs Jill Cross
330 Mr & Mrs M.J. Wright
331 Bessie Edith Sharps
332 Diana Oscroft
333 D.G. MacFarlane
334 Stuart McConnachie
335 Joyce Woodin
336 R.K.E. & L.E. Torode
337 Evelyn Fleming
338 G.W. Taylor
339 R.G.N. Bird
340 Dr A.N.B. Simpson
341 Dorothy J. Herlihy
342 E.E. Hawkeswood
343 Dr Michael E. Archer
344 John Raby
345 Roger A. Baulk
346 Shirley A. Rogers
347 Michael W. Woodgate
348 D.E.I. Wilks
349 P.E. Howell
350 Brian Yates
352 Andrea Peters
352 Dr John W. Everett
353 Harold H. Fowkes
354 Dr C.B. Wain
355 A.G. Millward
356
357 Dr Paul Foster
358 Audrey Diana Trumper
359 Roderick V. Perkins
360 Miss D.A. Phillips
361 Graham Davies
362 C.J. Ford
363 Margaret Howe
364 Tony Briggs
365 Arthur William Powell
366 A.J. Cottrill
367 M.C. Edwards
368 Joan M. Walker
369 J.R. Crane
370 Mrs F.M. Airey
371 J.S. White
372 Dr Alison Millward
373 Mrs Margaret M. Pontifex
374 Christopher John Lane
375 P.A. Eley
376 Dr D.J.L. Harding
377 W.G. Hardie
378 John Massey
379 W.J. Casey
380 Mrs Olive Belchamber
381 Sheila Harvey
382 Mrs Jill Winter
383
384 Andrew Dyke
385 Mike Davis
386 Mrs E.V. Davis
387 C. Crump
388 R.A. Ogden
389 L.H. & L.C. Doncaster
390 Brenda Jeanne Radbourne
391 E.W. Rose
392 Mrs Y. Turk
393 Jasmine R. Arnold
394 Dr M.B. Skirrow
395 T.N.D. Peet
396 Mr & Mrs W.G.G. Rollason
397 Anne Needler
398 Jean Kennedy
399 V.J. Barnes
400 S.A. Hammond
401 Miss M.E. Barling
402 W.H. Crowther
403 Phyllybeans
404 Robin Shaw
405 Mrs M.L. Williams
406 Kevin J. Rees
407 W.D. Trow
408 D.W. Trow
409 J.M. Moore
410 Marjorie E.A. Lloyd
411 Jackie Ann Pickett
412 Thomas A. Seymour
413 B.F. Savage
414 Anne Hale
415 E. Holloway
416 Linda Cooke
417 Robert K. Trumper
418 Miss Rosa Watson
419 Mike & Carol Priest
420 Les & Margaret Deaville
421 F. Fincher
422 David G. Coker
423 Jeffrey Perry
424 John Kendal
425 Mrs Barbara Warr
426 P.A. Jeavons
427 N.C.H. Pitt
428 E.B. Dennis
429 Dr Pam Gough
430 Georgina Coen
431 David H. Naylor
432 Paul Victor Capper
433 Mrs Joan M. Haden
434 Mr & Mrs P.T. Seaton
435 Mrs M.H. Brazil
436 Mrs A.M. Tomkins
437 Kevin Heede
438 Andrew John Ball
439 R.J. Price
440 John Preston Robinson
441 Richard Howell
442 Mrs P.E. Howell
443 D.R Smith
444 Daphne P. Maw
445 B.K. Taylor
446 Michael Hingley
447 Robert Charteris
448 Dr G. Ayliffe
449 Dr John V. Tranter
450 D.J. Triggs
451 Miss Audrey L. Clark
452 J.L.R. Ledbury
453 B. Marsh
454 Chaz Mason
455 Tabby & Frank Lucas
456 Julian G. Greenwood
457 Lorna J. Cameron
458 I.J. Wheeler
459 Brian Middleton
460 J.R. Blake
461
462 Robert W. Slater
463 Miss I. Birch
464 Bevan Craddock
465 Carol Gilbert
466 C.E. Wright
467 G.J. Willetts
468 P.F. Hollway
469 G.N. Allan-Smith
470 Mrs J. Heynes
471 Cambray Jones
472 Jeff Orgee
473 Kenneth Oddy
474 P.S. Lawless
475 Dr Ken Dickinson
476 Ian Gary Cave
477
478 Mrs Sue Wolfendale
479 Christiana Bentley
480 Dr D.W. James
481 Mike Bryan
482 David Steel
483 M.J. Buller
484 D.W. Scott
485 P.D. Rogers
486 Christopher John Lane
487 D.J. Simkin
488 Nan Waite
489 Alan Reavenall
490 T.B. Hutton
491 Simon Knight
492 H. & G. Bridges
493 P.J.M. Nethercott
494 Bromsgrove School
495 G.A. McCleverty
496 Marjorie Randall
497 Paul Moseley
498 Miss P.R. Jones
499 Tony Hartwell
500 Dr Marshall Wilson
501 Dr & Mrs David Wilson
502 Kate & Marc Young
503 Philip Ireland
504 Terence Paul Jenkins

505 William Weaver
506 Keith N. Kingscott
507 Gillian Coupe
508 Miss S. Matthews
509 Mr Ward
510 Mr & Mrs Grainger
511 Don & Myrtle Mather
512 Allan Ashford
513 John Tout
514 H. Mason
515 R.F. Simons
516 Mrs Marlane Newbury
517 S.R.J. Barker
518 J.B. Fowler
519 Mrs Janet S. Thom
520 T.O. Knight
521 Mrs Michael Godwin
522 Jacquiline Gill
523 K.N.A. Alexander
524 Roy M. Nicholls
525 R. Hartley-Cocks
526 R.M. Hatton
527 David G. Evans
528 R.V.J. Cadbury
529 Ann Okey
530 M.J. & L.L. Butler
531 Robert H. Carr
532 Robert Goodman
533 B.G. Bowen
534 Mrs S.V. Hill
535 Miss E. Phillips
536 John & Brenda Allan
537 Ann Thomas
538 G.D. Clemmow
539 Julie H. Clarke
540 Michael G. Needham
541 D.J. Cramp
542 Mark & Clare Kitchen
543 Brian Iles
544 Mrs C. Johnson
545 Margaret Tillott
546 C.A. Norris
547 A.L. Yoxall
548 A. Darby
549 P.L. Beacham
550 Mr & Mrs I.M. Ralphs
551 Miss E.J. Garlick
552 John V. Garrett
553 Mrs Margaret Lorton
554 Michael J. Bland
555 K.G. Clifford
556 David John Eaton
557 M.B. Nagington
558 Peter & Mary Fraser
559 Dr Anne Sutcliffe
560 W.R. Martin
561 Philip Bell
562 G.W. Drinkwater
563 Jill Harvey

Remaining names unlisted.

564 Dennis Barnes
565 Anne & Bill Adams
566 Sally Bullock
567 C.M. Hemmings
568 David Preece
569 Ray Atkins
570 Brian E. Hill
571 P.J. White
572 Julia Davies
573–574 Miss W.V. James
575 Martin Lewis
576 Dr & Mrs John Bartholomew
577 L. & W.J. Alexander
578 T.A. Parkes
579 Malcolm J. Warner
580 S.G.A. Roberts
581 Paul A. Janke
582 Christine Griffin
583 M.G. Everiss
584 Air Commodore Peter Hugues CBE DFC
585 Derek Gordon Fowle
586 Leila Manser
587 Arthur Reeves
588 Patricia S. Cooper
589 J.D. Andrews
590 D.J. Bradley
591 Dennis Jenkins
592 Zoe & Gordon Long
593 Mrs Dillis Smith
594 H.W. King
595 W.G. Gaman
596 Mr & Mrs J.E. Morris
597 Margaret Foden
598 F. Lancaster
599 Brian John William Watkins
600 T.M. Chew
601 W.D. Francis
602 Graham Wallhead
603 Trevor Trueman
604 L.A. Hoad
605 Miss N.M. Whitehead
606 Miss C.B. Fletcher
607 Miss A. Brayford
608 Gordon Raymond Kirk
609 C.H. Allcock
610 David Grahame Marsden
611 D. Medley
612 David Hughes
613 John Frazier
614 Mr & Mrs R. Dyson
615 Madeline M. Johnson
616 S.F.L. Edwards
617 R. Willcox

618 Mrs J.D. Parr
619 Victor Carvell
620–621 Jane Clark
622 Mrs J. Preston
623 P.V. Capper
624 Sheila Campbell
625 Martin Smith
626 Mr & Mrs C. Tucker
627 Mrs J.E. Chamberlayne
628 George S. Green
629 Norman Roberts
630 Mary Daulman
631 D.C. Boyce
632 Helen Gibson
633 Mrs Sherri James
634 I.R. Denton
635 J.K. Earle
636–638 Kathleen F. Meyer
639 Tim Hemming
640–642 John Graham McGregor-Smith
643 Martyn Stilliard
644 David R. Shackley
645 M.C. McCreedy
646 Brian Hadley
647 R.G. Stickland
648–650 Derek Woodward
651 R. Bubb
652 John Stanley
653 C.J. Washington
654 R.J. Mann
655 Mrs S.I. Brown
656 Brian G. Prudden
657 A.L. Norton
658 D.B. Honeyborne
659 Ann Parker
660 V. Mary Walker
661 Mrs Jennifer Brooks
662 M.H. Taylor
663 M.J. Hebden
664 Robert J. Gillmor
665 Liz Bomford
666 Muriel Harris
667 Frederick Mason
668 David Merry
669 David & Brenda Rea
670 Mrs C.M. Pettet
671 Kay Dellbridge
672 Mrs Christine P. Hawker
673 D.A. Coles
674 Mrs M. Watson
675 A.H. Fryer
676 Mrs B. Mayes
677 E.J. Lofthouse
678 H.A. Plumb

679 Miss E.R. New
680 B.H. & E.R. Adlam
681 Lucy Anne Bishop
682 Maureen Butcher
683 John Halstead
684 A.W. & L.A. Reid
685 David H. Malcomson
686 K.P. & J.A. Harman
687 Mrs Rose Hill
688 Terence Michael Jennings
689 C.D. Bancroft
690 The Baskeyfield Family
691 H.L. & C.E. Payne
692 Susan Finch
693 J. Gallé
694 Mrs V. Sinclair
695 R.W. Bradley
696 Mrs M. Shaw
697 Mrs A. Bright Clarke
698 Garth Lowe
699 James H. Toye
700 Unlisted
701 Nigel Fellows
702 G.H. Trevis
703 M.A. Moore
704 Norman Brown
705 Margaret Embleton
706 Ms B.M. Farrer
707 K. Crump
708–710 Cyril Hope OBE
711 R.G. Baylis
712 John & Diana Poole
713 Worcester City Library
714 Stourport Library
715 Redditch Library
716 Pershore Library
717 Malvern Library
718 Kidderminster Library
719 Droitwich Library
720 Evesham Library
721 Bromsgrove Library
722–774 Hereford & Worcester County Council
775 A.W. Cundall
776 P.J. Roberts
777 Grahame Wilcox
778 David & Rachel Poloni
779 Lee Scott
780–804 Worcestershire Nature Conservation Trust
805 Peter Davey
806 Mr & Mrs B. Westwood
807 Mr & Mrs R. Hill
808 S.W. Jones